How Your Brain Works

Inside the Most Complicated Object in the Known Universe

How Your Brain Works

*Inside the Most Complicated Object in
the Known Universe*

NEW SCIENTIST

**New
Scientist**

The publisher has used its best endeavours to ensure that any website addresses referred to
in this book are correct and active at the time of going to press. However, the publisher and
the author have no responsibility for the websites and can make no guarantee that a site
will remain live or that the content will remain relevant, decent or appropriate.
The publisher has made every effort to mark as such all words which it believes to be
trademarks. The publisher should also like to make it clear that the presence of a word in
the book, whether marked or unmarked, in no way affects its legal status as a trademark.
Every reasonable effort has been made by the publisher to trace the copyright holders
of material in this book. Any errors or omissions should be notified in writing to the
publisher, who will endeavour to rectify the situation for any reprints and future editions.
Cover image © Callista Images / Getty Images
Typeset by Cenveo® Publisher Services.
Printed and bound in the United States of America.
Nicholas Brealey Publishing policy is to use papers that are natural, renewable and recyclable
products and made from wood grown in sustainable forests. The logging and manufacturing
processes are expected to conform to the environmental regulations of the country of origin.

Nicholas Brealey Publishing
Carmelite House
50 Victoria Embankment
London EC4Y 0DZ
Tel: 020 7122 6000

Nicholas Brealey Publishing
Hachette Book Group
53 State Street
Boston, MA 02109, USA
Tel: (617) 523 3801

www.nicholasbrealey.com

**Also available
as an ebook**

Contents

Series introduction

New Scientist's Instant Expert books shine light on the subjects that we all wish we knew more about: topics that challenge, engage enquiring minds and open up a deeper understanding of the world around us. *Instant Expert* books are definitive and accessible entry points for curious readers who want to know how things work and why. Look out for the other titles in the series:

Scheduled for publication in spring 2017:
The End of Money
The Quantum World
Where the Universe Came From

Scheduled for publication in autumn 2017:
Evolution
Machines That Think
Our Beautiful Minds
Why The Universe Exists

Contributors

Editor-in-chief: Alison George is *Instant Expert* editor for *New Scientist*.

Editor: Caroline Williams is a UK-based science journalist and editor. She is author of *Override* (Scribe, 2017).

Articles in this book are based on talks at the 2016 *New Scientist* masterclass "How your brain works" and articles previously published in *New Scientist*.

Academic contributors

Daniel Bor is an author and cognitive neuroscientist at the Sackler Centre for Consciousness Science at the University of Sussex, UK.

Derk-Jan Dijk is professor of sleep and physiology at the University of Surrey in Guildford, UK and director of the Surrey sleep research centre.

Jonathan K. Foster is professor in clinical neuropsychology and behavioural neuroscience, affiliated to Curtin University in Perth, Australia, the neurosciences unit of the health department of Western Australia and the University of Western Australia.

Linda Gottfredson is professor emeritus of education at the University of Delaware in Newark and focuses on the social implications of intelligence.

Andrew Jackson is at the Institute of Neuroscience, Newcastle University, and is working on a neural prosthesis to restore hand movement after spinal injury, and on a brain implant to control epilepsy.

George Mather is professor of vision science at the University of Lincoln, UK. He specialises in the perception of movement and of visual art.

Michael O'Shea is professor of neuroscience and co-director of the centre for computational neuroscience and robotics at the University of Sussex, UK.

Tiffany Watt Smith is a research fellow at the centre for the history of the emotions at Queen Mary University of London.

Raphaëlle Winsky-Sommerer researches circadian rhythms and sleep at the University of Surrey in Guildford, UK.

Thanks also to the following writers:

Sally Adee, Anil Ananthaswamy, Colin Barras, Andy Coghlan, Catherine de Lange, Linda Geddes, Alison George, Jessica Griggs, Anna Gosline, Jessica Hamzelou, Bob Holmes, Courtney Humphries, Christian Jarret, Graham Lawton, Jessica Marshall, Alison Motluk, Helen Phillips, Michael Reilly, David Robson, Laura Spinney, Kayt Sukel, Helen Thomson, Sonia van Gilder Cooke, Kirsten Weir, Caroline Williams, Clare Wilson, Emma Young.

Introduction

If you are reading this, you are the proud owner of one of the most complex objects in the known universe: the human brain.

You wouldn't know this from looking at it: at first glance it's a 1.4 kilogram pinkish wrinkled blob with roughly the consistency of tofu. It looks so uninspiring, in fact, that until 2,500 years ago it was thought to do nothing more complex than cool the blood.

Now, of course, we know that the brain is a rich tangle of 86 billion neurons which, through a complex ballet of electrical and chemical activity, allows us to experience the world, feel, taste and remember. Over the course of human history it is what has enabled our species to build civilisations, create great art and fly to the moon.

The question of how it manages these feats has kept great minds busy for centuries. In recent decades, though, neuroscientists have had the distinct advantage of being able to use modern brain-imaging techniques to observe in real time as patterns of electrical activity and blood flow hint at what is going on inside.

As these techniques continue to reveal the brain's workings, neuroscience is forging into new territory, trying to piece together the entire wiring diagram of the human brain. It is, without doubt, the most exciting time in the history of brain science.

As we enter this exciting period of discovery, this *New Scientist Instant Expert* guide tells you everything you need to know about the human brain. Gathering together the thoughts of leading neuroscientists, and the very best of *New Scientist* magazine, it will bring you up to date with what the best brains in science know. If you have ever wondered how the brain senses, remembers, how it becomes conscious and what it is doing while we sleep, then read on.

Caroline Williams, Editor

I
Welcome to your brain

The brain is the most confusing, complicated and arguably the ugliest organ in our body — yet, in essence, it is simply a collection of nerve cells gathered together in one place to simplify the wiring. A brain can be just a handful of cells, as found in some simple invertebrates, or billions, as in humans. It allows animals to adapt their behaviour to changes in the environment on a much quicker timescale than evolution. Thanks to advances in the field of neuroscience, we now have an exquisite understanding of the brain's underlying architecture. But how did our human brains evolve, and what makes them different from those of other animals? And what are the philosophical implications of being "just a brain"? Here is a whistlestop tour of your grey matter.

The birth of neuroscience

The birth of neuroscience began with Hippocrates some 2,500 years ago. While his contemporaries, including Aristotle, believed that the mind was to be found in the heart, Hippocrates argued that the brain is the seat of thought, sensation, emotion and cognition.

It was a monumental step, but a deeper understanding of the brain took a long time to follow, with many early theories ignoring the solid brain tissue in favour of fluid-filled cavities, or ventricles. The 2nd-century physician Galen – perhaps the most notable proponent of this idea – believed the human brain to have three ventricles, with each one responsible for a different mental faculty: imagination, reason and memory. According to his theory, the brain controlled our body's activities by pumping fluid from the ventricles through nerves to other organs.

Such was Galen's authority that the idea cast a long shadow over our understanding of the brain, and fluid theories of the brain dominated until well into the 17th century. Even such luminaries as French philosopher René Descartes compared the brain to a hydraulically powered machine. Yet the idea had a major flaw: a fluid could not move quickly enough to explain the speed of our reactions.

A more enlightened approach came when a new generation of anatomists began depicting the structure of the brain with increasing accuracy. Prominent among them was the 17th-century English doctor Thomas Willis, who argued that the key to how the brain worked lay in the solid cerebral tissues, not the ventricles. Then, 100 years later, Italian scientists Luigi Galvani and Alessandro Volta showed that an external source of electricity could activate nerves and muscle. This was a crucial development, since it finally suggested why we respond so rapidly to events. But it was not until the 19th century that German physiologist Emil du Bois-Reymond confirmed that nerves and muscles themselves generate electrical impulses.

This paved the way for the modern era of neuroscience, beginning with the work of the Spanish anatomist Santiago Ramón y Cajal at the dawn of the 20th century. His spectacular observations identified neurons as the building blocks of the brain. He found them to have a diversity of forms that is not found in the cells of other organs. Most surprisingly, he noted that insect neurons matched and sometimes exceeded the complexity of human brain cells. This suggested that our abilities depend on the way neurons are connected, not on any special features of the cells themselves. Cajal's "connectionist" view opened the door to a new way of thinking about information processing in the brain, which still dominates today.

Wired to think

While investigating the anatomy of neurons in the 19th century, Santiago Ramón y Cajal proposed that signals flow through neurons in one direction. The cell body and its branched projections, known as **dendrites**, gather incoming information from other cells. Processed information is then transmitted along the neuron's long nerve fibre, called the **axon**, to the **synapse**, where the message is passed to the next neuron (*see* diagram, below).

It took until the 1940s and 50s for neuroscientists to get to grips with the finer details of this electrical signalling. We now know that the messages are transmitted as brief pulses called **action potentials**. They carry a small voltage – just 0.1 volts – and last only a few thousandths of a second, but they can travel great distances during that time, reaching speeds of 120 metres per second (m/s).

The nerve impulse's journey comes to an end when it hits a synapse, triggering the release of molecules called **neurotransmitters**, which carry the signal across the gap between neurons. Once they reach the other side, these molecules briefly flip electrical switches on the surface of the receiving neuron. This can either excite the neuron into sending its own signal, or it can temporarily inhibit its activity, making it less likely to fire in response to other incoming signals. Each is important for directing the flow of information that ultimately makes up our thoughts and feelings.

The complexity of the resulting network is staggering. We have around 86 billion neurons in our brains, each with around 1,000 synapses. If you started to count them at one per second you would still be counting 30 million years from now.

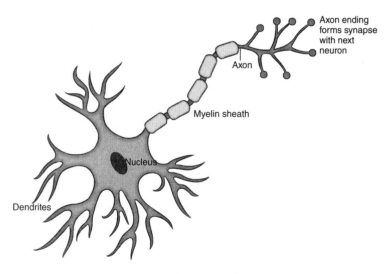

Axon ending forms synapse with next neuron

Axon

Myelin sheath

Nucleus

Dendrites

FIGURE 1.1 Structure of a neuron

Unlike the electronic components of a computer, our networks of neurons are flexible thanks to a special class of neurotransmitter. These **neuromodulators** act a bit like a volume control, altering the amount of other neurotransmitters released at the synapse and the degree to which neurons respond to incoming signals. Some of these changes help to fine-tune brain activity in response to immediate events, while others rewire the brain in the long term, which is thought to explain how memories are stored.

Many neuromodulators act on just a few neurons, but some can penetrate through large swathes of brain tissue creating sweeping changes. Nitric oxide, for example, is so small (it's the 10th smallest molecule in the known universe, in fact) that it can easily spread away from the neuron at its source. It alters receptive neurons by changing the amount of neurotransmitter released with each nerve impulse, kicking off the changes that are necessary for memory formation in the hippocampus.

Through the actions of a multitude of chemical transmitters and modulators, the brain is constantly changing, allowing us to learn, change and adapt to the world around us.

How did our brains get so complicated?

14 million years ago a small ape lived in Africa. It was a very smart ape but the brains of most of its descendants – orang-utans, gorillas and chimpanzees – do not appear to have changed greatly compared with the branch of its family that led to modern humans. What made us different?

We can only speculate as to why their brains began to grow bigger around 2.5 million years ago, but it is possible that serendipity played a part.

In other primates, the "bite" muscle exerts a strong force across the whole of the skull, constraining its growth. In our forebears, this muscle was weakened by a single mutation, perhaps opening the way for the skull to expand. This mutation occurred around the same time as the first hominids with weaker jaws and bigger skulls and brains appeared.

The development of tools to kill and butcher animals around 2 million years ago would have been essential for the expansion of the human brain, since meat is such a rich source of nutrients. A richer diet, in turn, would have opened the door to further brain growth.

Primatologist Richard Wrangham at Harvard University thinks that fire played a similar role by allowing us to get more nutrients from our food. Eating cooked food led to the shrinking of our guts, he suggests. Since gut tissue is expensive to grow and maintain, this loss would have freed up precious resources, again favouring further brain growth.

Our big brains might also have a lot to do with our complex social lives. If modern primates are anything to go by, our ancestors likely lived in groups. Mastering the social niceties of group living requires a lot of brain power. Robin Dunbar at the University of Oxford thinks this might explain the enormous expansion of the frontal regions of the primate neocortex, particularly in the apes. Dunbar has shown there is a strong relationship between the size of primate groups, the frequency of their interactions with one another and the size of the brain regions that deal with them.

Overall, it looks as if a virtuous cycle involving our diet, culture, technology, social relationships and genes led to the modern human brain coming into existence in Africa by about 200,000 years ago.

So where do we go from here? The answer is that we are still evolving. According to one recent study, the visual cortex has grown larger in people who migrated from Africa to northern latitudes, perhaps to help make up for the dimmer light up there.

Interestingly, we may have reached a point at which there is no advantage in our brains getting any bigger. There may have come a time in our recent evolutionary past when the advantages of bigger brains started to be outweighed by the dangers of giving birth to children with big heads. Or it might have been that our brains got too hungry to feed. Our brains already burn 20 per cent of our food intake and it could simply be that we can't afford to allocate any more energy to the job of thinking.

What's more, our brains might even be shrinking. In the past 10,000 years or so the average size of the human brain compared with our body has shrunk by 3 or 4 per cent. Some people wonder if it means we are getting stupider (*see* Chapter 3 for more on this). Others are more hopeful, suggesting that perhaps the brain's wiring is more efficient than it used to be.

Mapping the mind

The brain may be a tangle of neurons but it is anything but disorganised. As each brain develops before birth it organises itself into a characteristic shape that, details aside, looks much the same in all of us. There is more than one way to carve up something as complicated as this and different regions have a dizzying number of names and descriptions. At its simplest, though, the brain can be divided into three parts, each of which deals with a particular kind of processing.

Hindbrain

As its name suggests, the hindbrain is located at the base of the skull, just above the neck. Comparisons of different organisms suggest it was the first brain structure to have evolved, with its precursor emerging in the earliest vertebrates. In humans it is made up of three structures: the medulla oblongata, pons and cerebellum

The **medulla oblongata** is responsible for many of the automatic behaviours that keep us alive, such as breathing, regulating our heartbeat and swallowing. Significantly, its axons cross from one side of the brain to the other as they descend to the spinal cord, which explains why each side of the brain controls the opposite side of the body.

A little further up is the **pons**, which also controls vital functions such as breathing, heart rate, blood pressure and sleep. It also plays an important role in the control of facial expressions and in receiving information about the movements and orientation of the body in space.

The most prominent part of the hindbrain is the **cerebellum**, which has a very distinctive rippled surface with deep fissures. It is richly supplied with sensory information about the position and movements of the body and can encode and memorise the information needed to carry out complex fine-motor skills and movements. Recent research has also linked it with fine-tuning of our emotional and cognitive skills.

Midbrain

The midbrain plays a role in many of our physical actions. One of its central structures is the **substantia nigra**, so-called because it is a rich source of the neurotransmitter **dopamine**, which turns black in post-mortem tissue. Because dopamine is essential for the control of movement, the substantia nigra is said to "oil the wheels of motion". Dopamine is also the "reward" neurotransmitter and is necessary for many forms of learning, compulsive behaviour and addiction.

Other regions of the midbrain are concerned with hearing, visual information processing, the control of eye movements and the regulation of mood.

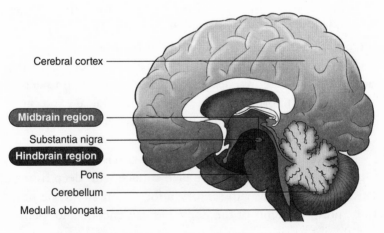

FIGURE 1.2 Identifying the major parts of the brain (and below)

Forebrain

Many of our uniquely human capabilities arise in the forebrain, which expanded rapidly during the evolution of our mammalian ancestors. It includes the **thalamus**, a relay station that directs sensory information to the **cerebral cortex** – the outer, wrinkly area of the brain – for higher processing; the **hypothalamus**, which releases hormones into the bloodstream for distribution to the rest of the body; the **amygdala**, which deals with emotion; and the **hippocampus**, which plays a major role in the formation of memories.

Among the most recently evolved parts are the **basal ganglia**, which regulates the speed and smoothness of intentional movements initiated by the cerebral cortex. Connections in this region are modulated by the neurotransmitter dopamine, provided by the midbrain's substantia nigra. A deficiency in this source is associated with many of the symptoms of Parkinson's disease, such as slowness of movement, tremor and impaired balance. Although drugs that boost levels of the neurotransmitter in the basal ganglia can help, a cure for Parkinson's is still out of reach.

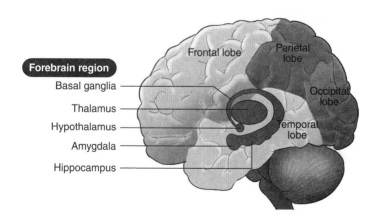

Finally, there is the cerebral cortex – the enveloping hemispheres thought to make us human. Here plans are made, words are put together and ideas generated. Home of our creative intelligence, imagination and consciousness, this is where the mind is formed.

Structurally, the cortex is a single sheet of tissue made up of six crinkled layers folded inside the skull; if it were spread flat it would stretch over 1.6 square metres. Information enters and leaves the cortex through about a million neurons, but it has more than 10 billion internal connections, meaning the cortex spends most of its time talking to itself.

Each of the cortical hemispheres have four principal lobes (*see* diagram). The **frontal lobes** house the neural circuits for thinking and planning, and are also thought to be responsible for our individual personalities. The **occipital** and **temporal lobes** are mainly concerned with the processing of visual and auditory information, respectively. Finally, the **parietal lobes** are involved in attention and the integration of sensory information.

The body in the brain

The body is "mapped" onto the cortex many times, including one map representing the senses and another controlling our movements. These maps tend to preserve the basic structure of the body, so that neurons processing feelings from your feet will be closer to those dealing with sensations from your legs than those crunching data from your nose, for example. But the proportions are distorted, with more brain tissue devoted to the hands and lips than the torso or legs. Redrawing the body to represent these maps results in grotesque figures like Penfield's homunculus (*below*).

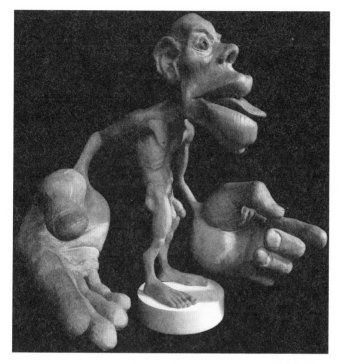

FIGURE 1.3 Penfield's homunculus: How the brain sees the body

A brain of two halves, and many, many parts

Brains come in two hemispheres, linked together by a tract of about a million axons, called the **corpus callosum**. Cutting this bridge, a procedure sometimes performed to alleviate epileptic seizures, can split the unitary manifestation of "self". It is as if the body is controlled by two independently thinking brains. One smoker who had the surgery reported that when he reached for a cigarette with his right hand, his left hand would snatch it and throw it away!

The corpus callosum allows different tasks, carried out by different cortical regions, to be combined smoothly into a seamless experience of the world. Objects are recognised with no awareness of the fragmented nature of the brain's efforts. Precisely how this is achieved remains a puzzle. It's called the "problem of binding" and is one of the many questions left to be answered by tomorrow's neuroscientists.

Mind readers: how scientists measure the brain and its activity

- **Magnetic resonance imaging (MRI)** Showing detailed anatomical images, it is like an X-ray for soft tissues.
- **Functional MRI (fMRI)** Displays changes in blood supply – assumed to correlate with local nerve activity – to different brain areas during mental tasks such as arithmetic or reading.
- **Diffusion MRI (also called diffusion imaging, tractography)** Reveals the brain's long-distance connections; works by tracking water molecules, which can diffuse along the length of axons more freely than escaping out through their fatty coating.
- **Functional connectivity MRI (resting-state MRI)** Also shedding light on long-distance connections, it measures spontaneous fluctuations in activity in different brain areas, which reveals the degree to which they communicate

---

A mind built on mathematics

While much has been learned from studies of particular brain areas and what they do, in recent years neuroscientists have begun to move away from describing the brain based on particular regions for particular skills, and are instead working towards understanding how the brain's network of neurons pulls these areas together to become more than a sum of their parts. This has spawned a new kind of neuroscience – a mathematics of the mind, which could reveal the very nature of human experience.

Small world big connections

If you stretched out all the nerve fibres in the brain, they would wrap four times round the globe. Crammed into the skull, you might think this wiring is a tangled mess, but in fact mathematicians know its structure well – it is a form of the "small-world network".

The hallmark of a small-world network is the relatively short path between any two nodes. You've probably already heard of the famous "six degrees of separation" between you and anyone else in the world, which reflects the small-world structure of human societies. The average number of steps between any two brain regions is similarly small, and slight variations in this interconnectivity have been linked to measures of intelligence.

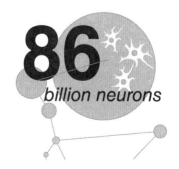

That may be because a small-world structure makes communication between different areas of a network rapid and efficient. Relatively few long-range connections

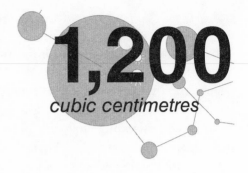

1,200
cubic centimetres

are involved – just 1 in 25 nerve fibres connect distant brain regions, while the rest join neurons in their immediate vicinity. Long nerve fibres are costly to build and maintain, says Martijn van den Heuvel at the University Medical Center in Utrecht, the Netherlands, so a small-world-network architecture may be the best compromise between the cost of these fibres and the efficiency of messaging.

The brain's long-range connections aren't distributed evenly over the brain, though. Van den Heuvel and Olaf Sporns of Indiana University at Bloomington recently discovered that clusters of these connections form a strong "backbone" that shuttles traffic between a dozen principal brain regions (*see* Figure 1.4). The backbone and these brain regions are together called a "rich club", reflecting the abundance of its interconnections.

No one knows why the brain is home to a rich club, says Van den Heuvel, but it is clearly important because it carries so much traffic. That makes any problems here potentially very serious. "There's an emerging idea that perhaps schizophrenia is really a problem with integrating information within these rich-club hubs," he says. Improving rich-club traffic flow might be the best form of treatment, though it is not easy to say how that might be achieved.

The brain's wiring allows for the rapid transmission of information, with a set of particularly well-connected hubs, known as the **rich club,** directing much of the traffic between different parts of the brain

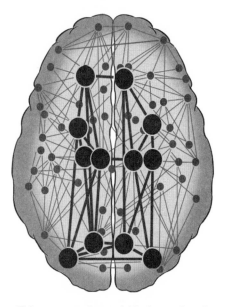

This group may be crucial for integrating all the thoughts and feelings that make up our conscious experience

FIGURE 1.4 The brain's 12 "rich-club" hubs

What is clear for now is that this highly interconnected network is the perfect platform for our mental gymnastics, and it forms a backdrop for

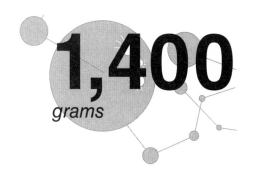

many of the other mathematical principles behind our thoughts and behaviour.

The edge of chaos

The brain's highly connected nature is undoubtedly useful, but it does have a potential downside. Because each neuron is linked into a highly connected small-world network, electrical signals can quickly spread far and wide, triggering a cascade of other cells to fire. So, theoretically, it could even snowball chaotically, potentially taking the brain offline in a seizure.

Thankfully, the chances of this happening are slight – only around 1 per cent of the population will experience a seizure in their lives. This suggests there is a healthy balance in the brain – it must inhibit neural signals enough to prevent a chaotic flood without stopping the traffic altogether.

An understanding of how the brain hits that sweet spot emerged in the 1970s, when Jack Cowan, now at the University of Chicago, realised that this balance represents a state known as the critical point or "the edge of chaos" that is well known to theoretical physicists. Cascades of firing neurons – or "neural avalanches" – are the moments when brain cells temporarily pass this critical point, before returning to the safe side, he said.

Avalanches, forest fires and earthquakes also result from systems lying at the critical point, and they all share certain mathematical characteristics. Chief among them is the so-called "power law" distribution, which means that

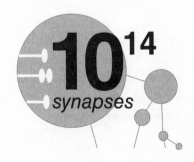

bigger earth-
quakes or for-
est fires happen
less often than
smaller ones
according to
a strict math-
ematical ratio;
an earthquake
that is 10 times
as strong as
another quake
is also just one-
tenth as likely to happen, for instance.

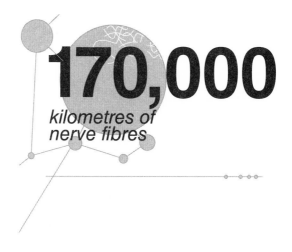

170,000 *kilometres of nerve fibres*

In 2003, John Beggs and Dietmar Plenz, both at the National Institute of Mental Health in Bethesda, Maryland, checked whether the brain also follows these rules. Sure enough, they found that, in rat brains, an excited neuron passed its signal to just one neighbour on average, which is exactly what you would expect of a system on the edge of chaos. Larger neural avalanches do occur, but, like earthquakes and forest fires, their frequency drops with size as predicted by a power law distribution.

Functional MRI scans have since suggested that the same kind of edge-of-chaos activity can be found at much larger scales, across the whole human brain; indeed, computer models suggest it might be a result of the small-world structure of the brain.

Balancing on the edge of chaos may seem risky, but the critical state is thought to give the brain maximum flexibility – speeding up the transmission of signals and allowing it to quickly coordinate its activity in the face of a changing situation. Some researchers are beginning to wonder whether certain disorders, like epilepsy, might arise when the brain veers away from this delicate balance. "Just as there's a healthy heart rate and a healthy blood pressure, this may be what you need for a healthy brain," says Beggs.

Ideas: fighting for survival

As your mind flits from thought to thought, it may seem as if dozens of sensations and ideas are constantly fighting for your attention. In fact, that's surprisingly close to the mark; the way different neural networks compete for dominance echoes the battle for survival between a predator species and its prey. Your wandering mind may be a by-product of the process.

Mikhail Rabinovich at the University of California in San Diego and Gilles Laurent at the California Institute of Technology in Pasadena, now at the Max Planck Institute of Brain Research in Frankfurt, Germany, first noticed that neuronal activation fluctuates in a kind of wave. They were studying neurons in the insect equivalent of the olfactory bulb, which processes odours. What they expected to see was something called habituation – that activity would drop off after the neuron had detected a smell. Instead, activity fluctuated up and down as time passed.

Looking closely, Rabinovich noticed that the pattern of activity looks suspiciously like the one described by mathematicians Alfred Lotka and Vito Volterra in the early 20th century to describe the interaction between predators and prey. These show that as a predator near-exhausts its supply of prey, the predator population begins to starve, allowing its prey to recover. When there are enough prey again, the cycle starts once more.

Rabinovich says something similar occurs in the brain. Here, though, the fight is not between just two competitors, but between the cognitive patterns that make up thought. None ever manages to gain more than a fleeting supremacy, which might explain the familiar experience of the wandering mind.

A better understanding of how these competitions play out in healthy brains and in conditions like attention-deficit hyperactivity disorder (ADHD) and obsessive compulsive disorder (OCD) might prove key to understanding how to nudge potentially unbalanced thought competitions back in a healthier direction.

Calculating the future

Another of the brain's biggest mathematical problems is making predictions from a crackling electrical storm of activity. Which words are likely to crop up next in a conversation, for example. Or whether a gap in the traffic is big enough to allow you to cross the road.

One explanation for how it does this comes from an area of mathematics known as **Bayesian statistics**. Named after 18th-century mathematician Thomas Bayes, the theory offers a way of calculating the probability of a future event based on what has gone before, while constantly updating the picture with new data. For decades, neuroscientists had speculated that the brain uses this principle to guide its predictions of the future, but Karl Friston at University College London took the idea one step further.

Friston looked specifically at the way the brain minimises the errors that can arise from these Bayesian predictions; in other words, how it avoids surprises. Realising that he could borrow the mathematics of thermodynamic systems like a steam engine to describe the way the brain achieves this, Friston called his theory "the free energy principle". Since prediction is so central to almost everything the brain does, he believes the principle could offer a general law for much, if not all, of our neural activity – the brain's equivalent of $E=mc^2$ in terms of its descriptive power and elegance.

So far, Friston has successfully used his free energy principle to describe the way neurons send signals backwards and forwards in the visual cortex in response to incoming sights. He believes the theory could also explain some of our physical actions. For instance, he has simulated our eye movements as we take in familiar or novel images, suggesting the way the brain builds up a picture with each sweep of our gaze to minimise any errors in its initial perception. In another paper he turned his attention to the delicate control of our arm as we reach for an object, using the free energy principle to describe how we update the muscle movements by combining internal signals from the turning joints with visual information.

Others are using the concept to explain some of the brain's more baffling behaviours. Dirk De Ridder at the University of Otago's Dunedin School of Medicine in New Zealand, for instance, has used the principle to explain the phantom pains and sounds people experience during sensory deprivation. He suggests they come from the neural processes at work as the brain casts about wildly to predict future events when there is little information to help guide its forecasts.

Interview: Am I really "just a brain"?

All of this talk of neurons, wires and mathematical laws can seem a little unnerving. Can our hopes, loves and very existence really be just the outcome of electricity moving through a mass of grey tissue? Neurophilosopher **Patricia Churchland** *says "yes", but the reality of it should inspire, not scare us.*

You compare revelations in neuroscience with the discoveries that the Earth goes around the Sun and that the heart is a pump. What do you think these ideas have in common?

They challenge a whole framework of assumptions about the way things are. For Christians, it was very important that the Earth was at the centre of the universe. Similarly, many people believed that the heart was somehow what made us human. And it turned out it was just a pump made of meat.

I think the same is true about realising that when we're conscious, when we make decisions, when we go to sleep, when we get angry, when we're fearful, these are just functions of the physical brain. Coming to terms with the neural basis of who we are can be very unnerving. It has been called "neuroexistentialism", which really captures the essence of it. We're not in the habit of thinking about ourselves that way.

Why is it so difficult for us to see the reality of what we actually are?

Part of the answer has to do with the evolution of nervous systems. Is there any reason for a brain to know about itself? We can get along without knowing, just as we can get along without knowing that the liver is in there filtering out toxins. The wonderful thing, of course, is that science allows us to know.

Are there any implications of neuroscience that you feel unsettled by?

I'd have to say no. It takes some getting used to, but I'm not freaked out by it. I certainly understand the ambivalence people have. On one hand, they're fascinated because it helps explain their mother's Alzheimer's but, on the other, they think "Gosh, the love that I feel for my child is really just neural chemistry?" Well, actually, yes it is. But that doesn't bother me.

You seem to take it for granted that there is resistance to brain science out there. What led you to that conclusion?

For many years I taught philosophy of neuroscience and my students would often say, doesn't it freak you out that you're just your brain? Doesn't that bother you? So we would talk about why it bothered them. I know some people are ambivalent or apprehensive.

You accept that we don't have satisfying neural explanations for a lot of higher functions, including consciousness, sleep and dreaming. Are we really ready to declare that we are our brains?

True, we don't have adequate explanations yet, and it's important not to overstate where things are. But that's where the evidence is pointing. Everything we're learning about in neuroscience points us in that direction.

You say beliefs in things like the existence of the soul and life after death are challenged by neuroscience. But are they still widely held?

There are probably cultural variations; it may be that in Britain there is less need to challenge these ideas. But I find that here, in America, it is important. Lots of people who don't necessarily have strong religious views nonetheless have the feeling that maybe after they die, there's something else.

Even people who have largely come to terms with neuroscience find certain ideas troubling – particularly free will. Do we have it?

A better question is whether we have self-control, and it's very easy to see what the evolutionary rationale of that is. We need to be able to maintain a goal despite distractions. We need to suppress certain kinds of impulses. We do know a little bit about the neurobiology of self-control, and there is no doubt that brains exhibit self-control.

Now, that's as good as it gets, in my view. When we need to make a decision about something – whether to buy a new car, say – self-control mechanisms work in ways that we understand: we decide not to spend more than we can afford, to go with the more or less practical car. That is what free will is. But if you think that free will is creating the decision, with no causal background, there isn't that.

Can neuroscience offer a philosophy to live by?

Neuroscience doesn't provide a story about how to live a life. But I think that understanding something about the nature of the brain encourages us to be sensible.

Some might say the idea that you are just your brain makes life bleak, unforgiving and ultimately futile. How do you respond to that?

It's not at all bleak. I don't see how the existence of a god or a soul confers any meaning on my life. How does that work, exactly? Nobody has ever given an adequate answer. My life is meaningful because I have family, meaningful work, because I love to play, I have dogs. That's what makes my life meaningful, and I think that's true for most people.

Now, at the end of it, what's going to happen? I will die and that's it. And I like that idea, in a crazy sort of way.

Patricia Churchland is a philosopher at the University of California, San Diego. She focuses on the intersection of neuroscience, psychology and philosophy.

What is special about the human brain, compared to the brains of other primates?

Human brains, particularly the cerebral hemispheres, are bigger and better developed than those in other primates. The frontal and prefrontal lobes of the cerebral hemispheres deal with complex kinds of thought and social interaction, such as planning, decision making, empathising, lying, and making moral judgements. But corrected for body size, the differences are surprisingly small.

The difference between a human brain and a chimp or gorilla brain seems to be largely down to the way neurons are connected. Humans have several unique genes that seem to control nerve cell migration as the brain develops and different patterns of gene expression in the brain.

So, the machinery doesn't look that different, but it certainly works differently.

As for non-primates – other mammals have smaller brains, with less well developed frontal lobes. Further down the evolutionary tree, animals lose the cortex altogether, with reptiles having a brain that resembles just our own brainstem. In simple animals, the brain becomes more of a swelling at the top of the nerve cord or around the mouth area.

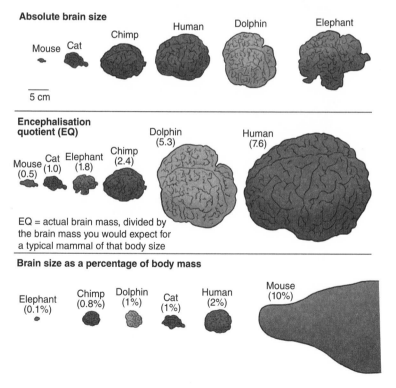

Inferring intelligence from brain size is questionable, not least because the relative sizes of brains change dramatically depending on how they are measured

Absolute brain size

Mouse Cat Chimp Human Dolphin Elephant

5 cm

Encephalisation quotient (EQ)

Mouse (0.5) Cat (1.0) Elephant (1.8) Chimp (2.4) Dolphin (5.3) Human (7.6)

EQ = actual brain mass, divided by the brain mass you would expect for a typical mammal of that body size

Brain size as a percentage of body mass

Elephant (0.1%) Chimp (0.8%) Dolphin (1%) Cat (1%) Human (2%) Mouse (10%)

FIGURE 1.5 Size isn't everything

How your brain evolved

Our brains followed a twisted path of development through creatures
that swam, crawled and walked the Earth long before we did.

850 million years ago
Animals first evolved. They had no
brains, but eventually evolved nerve
cells that could communicate with each
other using electrical pulses and
chemical signals.

600 million years ago
Neurons evolved when nerve cells
developed long-extensions – axons – to
carry electrical signals over
long distances.

6 million years ago
The human lineage diverges
from that of its closest relatives:
chimpanzees and bonobos.

14 million years ago
The ape that gave rise to us was living in Africa.
The brains of most of its descendants –
orang-utans, gorillas and chimpanzees –
have not changed much compared
with the branch that led to us.

2.5 million years ago
Homo habilis, the first of our genus
Homo evolves. Its brain is at least 30
per cent bigger than that of the more
ancient ape-like *Australopithecine*
hominids

1.8 million years ago
Homo erectus evolved, with a brain
roughly 50 per cent bigger than that of
Homo habilis.

550 million years ago
Central nervous systems developed around this time, allowing information to be processed rather than just relayed.

500 million years ago
Small brain-like structures are present in primitive fish-like creatures. Later, many core structures still found in our brains evolved.

55 million years ago
Primates evolved. The challenge of keeping track of their social lives might explain the enormous expansion of the frontal regions of the primate neocortex.

200 million years ago
The first mammals evolved. These creatures had a small neocortex – layers of neural tissue on the brain surface responsible for the complexity and flexibility of mammalian behaviour.

600,000 years ago
Homo Heidelbergensis lives in Africa and Europe, the shared ancestor of modern humans and Neanderthals. It has a similar brain capacity to us.

200,000 years ago
Modern humans came into existence in Africa, along with our modern brain – which was more than double the size of that of *Homo habilis*.

10,000 years ago
The average size of the human brain begins to shrink. It has now declined by 3 to 4 per cent relative to body size.

2
Memory

The capacity to remember the past is an integral part of human exist-ence. Without it, you would not be able to drive to work, hold a mean-ingful conversation with your family, read a book, or prepare a meal. Understanding what memory is and how it works is a fundamental goal of modern neuroscience.

What is memory?

Plato famously compared our memory to a wax tablet that is blank at birth and slowly takes on the impression of the events from our life. Only in the past hundred years, though, have psychologists developed objective techniques to study our recollections of the past. What has become clear is that human memory is a lot more complicated than Plato imagined.

For a start, it is not just one thing, but many. Scientists use three subtypes of memory storage – sensory, short term and long term – to describe how long different kinds of information hang around in the brain.

Sensory memory

During every moment of an organism's life, its eyes, ears and other sensory organs are taking in information and relaying it to the nervous system for processing. Our sensory memory store retains this information for a few moments. So twirling a sparkler, for example, allows us to write letters and make circles in the air thanks to the fleeting impression of its path.

Johann Segner, an 18th-century Hungarian scientist, was one of the first to explore this phenomenon. He reportedly attached a glowing coal to a cartwheel, which he rotated at increasing speeds until an unbroken circle of light could be perceived. His observations were followed by the systematic investigations of US psychologist George Sperling 200 years later. By studying people's ability to recall an array of letters flashed briefly on a screen, he found that our fleeting visual impressions – dubbed "iconic memory" – last for just a few hundred milliseconds. Studies of "echoic" sound memories came soon afterwards, showing that we retain an impression of what we hear for several seconds. Of note, echoic memories may be impaired in children who are late talkers.

Sensory memories are thought to be stored as transient patterns of electrical activity in the sensory and perceptual regions of the brain. When this activity dissipates, the memory usually fades too. While they last, though, they provide a detailed representation of the entire sensory experience, from which relevant pieces of information can be extracted into short-term memory and processed further via working memory.

Short-term and working memory

When you hold a restaurant's phone number in your mind as you dial the number, you rely on your short-term memory. This store is capable of holding roughly seven items of information for approximately 15 to 20 seconds, though actively "rehearsing" the information by repeating it several times can help you to retain it for longer.

Seven items of information may not seem much, but it is possible to get around this limit by "chunking" larger pieces of information into meaningful units. To recall a ten-digit telephone number, for instance, a person could chunk the digits into three groups: the area code (such as 021), then a three-digit chunk (639) and a four-digit chunk (4345).

Your short-term memory seems to store verbal and visuospatial information in different places. The verbal store has received most attention. Its existence has been inferred from studies asking volunteers to remember lists of words: people tend to be much better at recalling the last few items in a list, but this effect disappears if the test is delayed by a few seconds, especially if the delay involves a verbal activity that interferes with the storage process, such as counting backwards.

Verbal short-term memories seem to be stored in acoustic or phonological form. When you try to remember sequences of letters, for instance, lists of letters that are similar in sound, like P, D, B, V, C and T, are harder to recall correctly than sequences of dissimilar-sounding letters like W, K, L, Y, R and Z, even when the information is initially presented visually.

Short-term memory is closely linked to **working memory**, and the two terms are often used interchangeably. There is a difference, however: short-term memory refers to the passive storage and recall of information from the immediate past, whereas working memory refers to the active processes involved in manipulating this information. Your short-term memory might

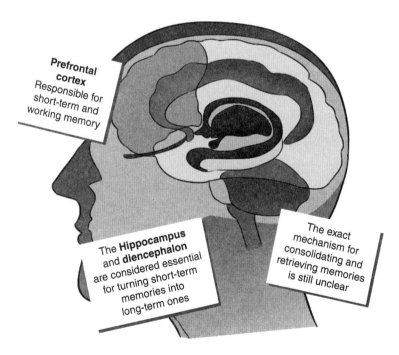

Prefrontal cortex
Responsible for short-term and working memory

The **Hippocampus** and **diencephalon** are considered essential for turning short-term memories into long-term ones

The exact mechanism for consolidating and retrieving memories is still unclear

FIGURE 2.1 Memory: from sense to storage

help you to remember what someone has just said to you, for example, but your working memory would allow you to recite it to them backwards or pick out the first letter of each word.

Long-term memory

Particularly salient information gets transferred to the brain's long-term storage facility, where it can remain for years or even decades. Your date of birth, phone number, car registration number and your mother's maiden name are all held here.

Unlike short-term memory, with its acoustic representations, we seem to store long-term memories by their meaning. If you try to recall information after a delay, for instance, you probably won't be able to reproduce the exact wording but its meaning or gist should come back fairly easily. This can lead to errors, however.

Long-term memories can take many different forms. **Semantic memories**, for example, concern your knowledge of facts, such as Paris being the capital of France, though you may not

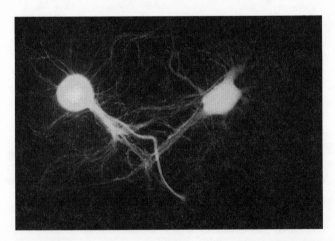

FIGURE 2.2 A single concept may be represented by 1 million neurons in your hippocampus (though these belong to a sea slug).

remember the exact circumstances in which you acquired this information. **Episodic memories** concern particular events from your life, such as the day you passed your driving test.

You can also categorise long-term memories by the way they influence your behaviour. Consciously recalled events or pieces of information are known as **explicit memories**, whereas **implicit memory** refers to experiences that influence your behaviour, feelings or thoughts without you actively recollecting the events or facts. For instance, if you pass an Italian restaurant on the way to work in the morning, you might later that day think about going out for an Italian meal, without being aware that you had been influenced by your morning journey.

From early influential work by Canadian psychologist Donald Hebb in the 1940s, through to the more recent Nobel-prize-winning work of US neuropsychiatrist Eric Kandel of Columbia University in New York, we now know that long-term memories are maintained by stable and permanent changes in neural connections. Early studies in sea slugs – creatures which can grow up to a foot long and have giant nerve cells to match – made it possible to watch what happens when a new memory forms.

These studies revealed that when sea slugs learned a simple response to a stimulus some of their synapses were strengthened. An impulse in the first neuron was now more likely to trigger the second to fire. This turns out to be the basis of memory in any animal with a nervous system. The work was so pivotal it earned Kandel the Nobel Prize in Physiology or Medicine in 2000.

More recently, with techniques such as fMRI, we have developed the capacity to study these processes non-invasively in humans. The **diencephalon** and the **hippocampus** regions seem to be essential for consolidating information from short-term into long-term memories.

What does a memory look like in the brain?

In the Harry Potter films, they are silver streams that can be teased from the head with the tip of a wand. In the Pixar movie *Inside Out*, they are small glowing balls, stored in vast racks of shelving in our minds. But what does a memory really look like? How does your brain store information for later retrieval? Where are memories stored and, if you could see them, what would they look like?

Answers are surprisingly hard to come by, in part because there are so many pieces of the memory puzzle. Some researchers focus on the minute details of what goes on in connections between brain cells. Others try to understand the subjective experience of memory – such as how, for Marcel Proust, the taste of a madeleine cake invoked detailed scenes from his childhood. However, the bigger picture of how the brain changes when we create a new memory is much more difficult to unravel.

Clues, though, are starting to come together into a fascinating glimpse into how memories form the glue that holds our lives and personalities together.

Kandel's work with sea slugs revealed the 'how' of memory formation, but it didn't answer the question of where the magic was happening.

The strange case of Henry Molaison

A major step along the route to answering this question came from one of the saddest tales of modern neuroscience. In 1953, Henry Molaison – known for a long time only by his initials, H.M. – had an operation that went badly wrong. The surgeon had been trying to remove brain tissue that was causing his epilepsy. Molaison's seizures originated in the hippocampi, a pair of structures either side of the brain, whose Latin name reflects their supposed resemblance to a seahorse. So the surgeon took them out.

The consequences for the 27-year-old Molaison were immense. Unable to hold a thought in his head for long, he needed care for the rest of his life. But equally profound was the impact on neuroscience – we have learned volumes from the way the surgery destroyed some of Molaison's abilities but spared others.

Molaison seemed to retain most of what he knew before the operation, suggesting that while the hippocampi are crucial for forming new memories, they are less important for storage. His short-term memory was unaffected, too – he could retain information for 15 to 30 seconds but no longer. In addition, Molaison's brain damage revealed that there are some important subdivisions of long-term memory. He could still learn physical skills, like riding a bike. But he had problems forming new memories of things that happened to him and in learning new facts.

The long and short of it

The hippocampi seem to be crucial, then, for the sort of memories that are central to our personal and intellectual lives. But they are by no means the only part of the brain required to make a memory. While the hippocampi are important and dominate current research, memory also involves the **cortex**, the outer layer of the brain that handles our complex thoughts and sensory perceptions of the world.

Say that yesterday you saw a rose in your garden and stopped to inhale its fragrance. This event was processed by specific parts of your cortex at the brain's back and sides that are responsible for vision and smell. Today, if you recall the experience, those same regions will be reactivated. This idea, sometimes known as **reinstatement**, has been around for a while, but was only confirmed in the past decade or so, thanks to the development

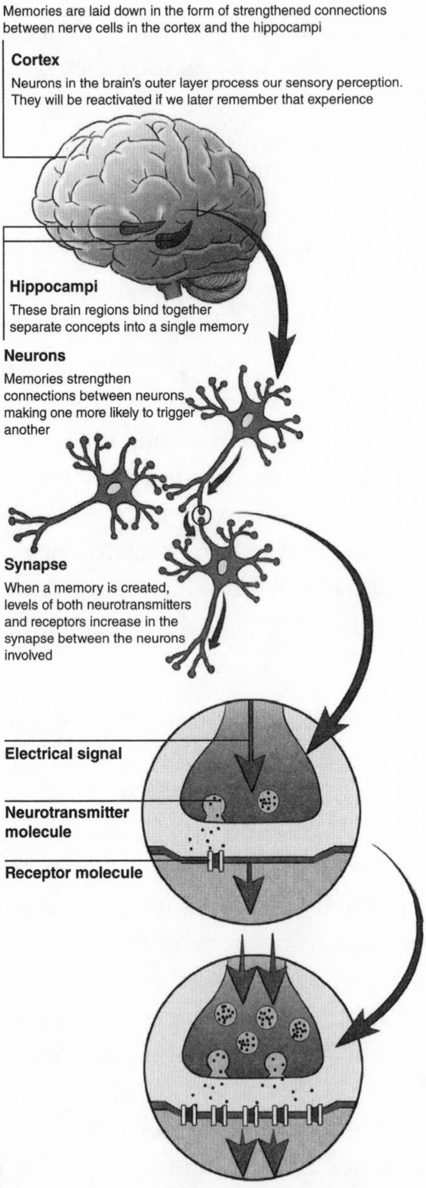

Memories are laid down in the form of strengthened connections
between nerve cells in the cortex and the hippocampi

Cortex

Neurons in the brain's outer layer process our sensory perception.
They will be reactivated if we later remember that experience

Hippocampi

These brain regions bind together
separate concepts into a single memory

Neurons

Memories strengthen
connections between neurons,
making one more likely to trigger
another

Synapse

When a memory is created,
levels of both neurotransmitters
and receptors increase in the
synapse between the neurons
involved

Electrical signal

**Neurotransmitter
molecule**

Receptor molecule

FIGURE 2.3 You must remember this

of brain scanning techniques. The same areas of the cortex light up in scanners both when someone first sees a picture of something and then is later asked to remember it.

A short-term memory of sniffing the rose wouldn't involve the hippocampi, as Molaison showed. But if, for some reason, you created a memory that lasted more than half a minute, then connections between the relevant areas of your cortex and your hippocampi would become strengthened. The hippocampi are wired up to many different parts of the cortex and help to combine the different aspects of a single memory.

This ability helps explain one of memory's hallmarks – that recalling one aspect of an experience can bring its other features to mind unbidden. Hearing a song on the radio can remind us of the moment we first heard it, for example, or a long-forgotten taste of madeleine cakes can remind you of childhood.

Memories, it seems, are made as a result of a spider's web of neurons firing together because of shared, strong connections. Strands of the web reach across different parts of the cortex and deep down to the hippocampi, the guardians of our memory bank.

Memory for the future

Given the infallibility of human memory, researchers are starting to wonder if it didn't evolve so that we could remember at all, but to allow us to imagine what might happen in the future. This idea began with the work of Endel Tulving, now at the Rotman Research Institute in Toronto, Canada, who discovered a person with amnesia who could remember facts but not episodic memories relating to past events in his life. Crucially, whenever Tulving asked him about his plans for that evening, the next day or the summer, his mind went blank – leading

Tulving to suspect that foresight was the flipside of episodic memory.

Subsequent brain scans supported the idea, suggesting that every time we think about a possible future, we tear up the pages of our autobiographies and stitch together the fragments into a montage that represents the new scenario. This process is the key to foresight and ingenuity, but it comes at the cost of accuracy, as our recollections become frayed and shuffled along the way. "It's not surprising that we confuse memories and imagination, considering that they share so many processes," says Daniel Schacter, a psychologist at Harvard University.

Why we forget

To function properly we need our memory to do three things:

- encode information in a storable form
- retain that information, and
- enable it to be accessed at a later point.

A failure in any of these components leads us to forget.

At the encoding stage, distraction or reduced attention can cause a memory failure. When already in storage memories can also "fade" and become less distinctive if the storage of other memories interferes with them, perhaps because they are stored in overlapping neural assemblies. Brain injury, too, can cut off the links between the hippocampus and the cortex that are necessary to recall information.

A failed memory search, such as having someone's name on the tip of your tongue, happens because the brain's search algorithms aren't perfect, and it may sometimes have trouble distinguishing the right signals from other neural noise. Many memory failures occur at the retrieval stage, when memories are re-encoded and are particularly vulnerable to being changed or lost.

Other things being equal, we tend to remember things that are the most important, such as information that is potentially rewarding or threatening.

The most extreme example of this is **flashbulb memory**, when strong memories are burned onto the brain in detail and last for years afterwards. They usually happen when an event is unusual, arousing or associated with strong emotions. Flashbulb memory is why so many people can remember where they were when they heard about the assassination of President John F. Kennedy in 1963, the death of Princess Diana in 1997, or the events of 9/11.

Another common phenomenon, known as the **reminiscence bump**, refers to the wealth of memories that we form and store between adolescence and early adulthood. When we are older, we are more likely to remember events from this period than any other stage of life, before or after. It could be that the reminiscence bump is due to the particular emotional significance of events that occur during that period, such as meeting one's partner, getting married or becoming a parent, and events that are life-defining in other ways, such as starting work, graduating from university or backpacking around the world.

Imagined memories

Strangely, some of the things we remember may not have happened at all. Memory is also vulnerable to being changed after the event. For example, if someone witnesses a traffic accident and is later asked whether the car stopped before or after the tree, they are much more likely to "insert" a tree into their memory of the scene, even if there was no tree present. This occurrence reflects the fact that when we retrieve a memory, we also re-encode it, and during that process it is possible to implant errors.

Elizabeth Loftus at the University of California, Irvine, and colleagues have shown that this "misinformation effect" could have huge implications for the court room, with experiments repeatedly demonstrating that eyewitness testimonies can be distorted by misleading questioning. Fortunately, their findings also suggest ways for police, lawyers and judges to frame the questions that they ask in a way that makes reliable answers more likely.

Related to the misinformation effect are "recovered" and false memories. A team led by Henry Roediger and Kathleen McDermott at Washington University in St Louis, Missouri, has built an extensive body of research showing that false memories can be formed relatively easily. People can be encouraged to "remember" an item that is linked in its meaning to a series of previously presented items but which itself was not presented. Suggestions and misleading information can create "memories" of personal events that the individual strongly believes to have happened in their past but which never took place.

In one famous experiment, Loftus persuaded subjects that they had seen Bugs Bunny at Disneyland, despite the fact that he's a Warner Bros character. Such findings may represent a serious concern for legal cases in which adults undergoing therapy believe that they have recovered memories of abuse in childhood. It could also mean that some of our cherished memories from our lives never happened the way we remember them at all.

What is it like to have a perfect memory?

Bob Petrella has what is termed highly superior autobiographical memory. He describes how it feels to remember almost every day of his life as though it were yesterday.

"When I recall my past, it's like watching a home movie. I feel exactly how I felt. I'll even feel the weather – if it was hot and sticky I'll remember how tight my clothes were and what I was wearing. All of my senses are triggered. I'll remember who I was with, what I was thinking, my attitude – everything is stirred in my imagination.

"I first noticed it when I was in high school. I would talk to my buddies about something that had happened when we were kids. I'd say 'yeah, remember, it was on 4 February, it was a Friday'. That's when I started realising that my memory was a little different.

"People misunderstand it. They think you've got a photographic memory; they call you Rain Man. But I don't have autism, or use any mind tricks. I haven't got a great memory for anything other than my past.

"I remember the bad stuff as well as the good. But one of the benefits of remembering your mistakes is that you can learn from them more easily than others. Being able to feel the way you felt when you previously made a mistake allows you to think 'OK wow, I won't do that again' in certain situations. Most of the time the bad days aren't really that bad, so I don't dwell on it – I like being in the present.

"One of the best things about being able to remember so much of my past is being able to think about times I spent with the people I love. I can go back to any time in their life that I was with them and remember it like it was yesterday. Then if they're no longer with me, it's like I can still spend time with them."

The surprising benefits of forgetting

Efficient forgetting is a crucial part of having a fully functioning memory. We forget because the brain has developed strategies to weed out irrelevant or out-of-date information (*see* "The need to forget", below). When we forget something useful, it could be that this pruning system is working a little too well.

It is a strategy whereby we discard information that is out of date – an old phone number or what we ate last week, for example. Since retrieving and using information solidifies it in memory, our mind gambles that the information we rarely retrieve is safe to discard.

Another problem is **absent-mindedness** where, for example, we fail to properly encode information about where we put our keys because our attention is elsewhere. Yet another problem is **blocking**, where the brain holds back one memory in favour of a competing memory, so we don't get muddled, for example, where a single word has two different meanings. Occasionally we retrieve the one we don't want first, then struggle to remember the other (*see* "Test your ability to forget").

Each of these strategies has an adaptive purpose, preventing us from storing mundane, confusing or out-of-date memories. We want to remember our current phone number, not an old one, and where we parked the car today, not last week.

Too much forgetting, as seen in amnesia, can be seriously disabling, but as Diane Van Deren demonstrates, it can have some upsides. Van Deren is one of the world's elite ultra-runners. In one recent race she ran more than 1,500 kilometres over 22 days. On some of those days, she ran for as long as 20 hours. Van Deren had always been good at sport, but her incredible endurance seems to be down in part to her poor short-term memory, again the result of brain surgery for epilepsy.

Each of the words below has a verb meaning and a different noun meaning which is more commonly used. For each of them, quickly try to come up with a word association for the verb meaning. For example, for DUCK, write "crouch".

● LOAF	_____	● SHED	_____
● POST	_____	● FENCE	_____
● COURT	_____	● LOBBY	_____
● ROOT	_____	● STUMP	_____
● SOCK	_____	● FAWN	_____
● LODGE	_____	● PRUNE	_____
● SIGN	_____	● DUCK	_crouch_____
● BARK	_____	● RAIL	_____
● PINE	_____	● SINK	_____
● BOWL	_____	● RING	_____

Most people find it difficult to temporarily "forget" associations with the more dominant noun meaning, and want to write "quack" next to DUCK

FIGURE 2.4 Testing the ability to suppress memories

Often, she just cannot remember how long she has been running for, underestimating the time by as much as 8 hours. "Most people with amnesia suffer a tyranny of the present," says Adam Zeman, a neurologist studying memory and epilepsy at the University of Exeter, UK, but Van Deren's inability to remember how long she has been running seems to free her from the feelings of fatigue that plague other runners. Perhaps, while others get caught up in the details of where they have been and where they are going, Van Deren gets into a more zen-like state that lets her run for longer without feeling so much strain. Of course, it could also be that after the challenges in her life Van Deren has a higher threshold for discomfort than most people.

For the rest of us, losing track of time on a long run is difficult, but there are certainly ways in which these findings affect us all. Suzanne Higgs, and colleagues at the University of Birmingham, UK, has found that simple distractions such as watching TV can stop people from forming good memories of what they are eating. As a result, they tend to snack more after the meal than control groups who were not distracted.

Imagination can play a powerful role too. Thanks perhaps to its close link to memory, simply imagining the process of eating something can lead people to feel more satiated, causing them to eat less. Which all goes to show that in the fight against overeating, working your memory could be your biggest ally.

Is my memory normal?

Some people can dredge up obscure details from conversations long ago, but not the places they have visited or the names of pop bands. Others are great at remembering facts, but all at sea when it comes to details of past personal experiences. Many people think they have less-than-stellar memory skills, but how much should we worry about the things we forget?

Perhaps surprisingly, the cleverest thing about memory is not what we remember, but what we forget. All brains discard most of the sensory data they receive because it isn't relevant to what happened. "Tomorrow you'll remember reasonably well a conversation you had today," says neurobiologist James McGaugh of the University of California, Irvine. "Within a week, a lot of that information will have been lost." Within a year, the conversation might be gone entirely.

Different types of memories hang around in the brain for different amounts of time. Sensory memories only last a few moments but some go on to make short-term memories, such as the phone number you just dialled. Exact figures are hard to pin down, but an average brain can probably keep around four things in mind at once, for up to 30 seconds.

Only really important or meaningful information makes it into long-term memory, such as a conversation that contained a personal insult. "We have selectively strong memories for events that are emotionally arousing," says McGaugh. Long-term memories divide into two main types. Semantic memories record facts, such as the concept of a train. Episodic ones are about events we have experienced, such as a particular train journey.

We probably all know someone who has an encyclopaedic factual memory, but extraordinary episodic memories are a more recent discovery. "These people remember events from years ago the way you and I remember events from last week," says McGaugh. This condition is called **highly superior autobiographical memory**, and there's also the opposite condition, in which people struggle to recall even recent events they have experienced. "They know the event happened, but they can't mentally travel back, even one week," says Daniela Palombo, who researches autobiographical memory at Boston University in Massachusetts (*see also* Petrella account above).

Most of us fall between these two extremes. True to the stereotype, women tend to have better episodic memories. With semantic memory, men tend to remember spatial information better, whereas women generally perform better at verbal tasks, such as recalling word lists. Personality type seems to be a factor, too: people open to new experiences tend to have better autobiographical memory.

Ageing affects the recall of personal experiences more than that of facts, as does depression. But if by our 40s we notice we can't remember new names, it's not that our brains are overloaded, since our memory capacity is practically unlimited. Rather, gradual changes in brain structure, such as a reduction in the density of dendrites that help to form connections between neurons, make the creation and retrieval of memories less efficient.

But until you start finding it difficult to carry out a simple task you have done many times before, or follow the flow of a conversation, you shouldn't be overly concerned if your memory seems to move in mysterious ways. Ultimately, memory is a personal thing, and no two brains do it the same way.

Six tips to master your memory

1 Hit the sweet spot

When trying to memorise new material, you may find yourself staring endlessly at the page in the hope that its contents will somehow seep into your mental vault. One of the most effective ways of learning for an exam, though, is to test yourself repeatedly, which may be simpler to apply to your studies than other, more intricate methods, such as the formal mnemonic techniques used by expert memorisers.

It's important to pace yourself, too, by revisiting material rather than cramming it all in during a single session. When doing so, you should make the most of sweet spots in the timing of your revision. If you are studying for an exam in a week's time, for instance, you will remember more if you leave a day or so between your first and second passes through the material. For a test in six months, revision should come about a month into your studies.

2 Limber up

Besides keeping your body – and therefore your grey matter – in generally good shape, a bit of exercise can offer immediate benefits for anyone trying to learn new material. In one study, students taking a ten-minute walk found it much easier to learn a list of 30 nouns, compared with those who sat around, perhaps because it helped increase mental alertness.

Short, intense bursts of exercise may be the most effective. In a recent experiment, participants learning new vocabulary performed better if their studies came after 2 × 3-minute runs, as opposed to a single 40-minute gentle jog. The exercise seemed to encourage the release of neurotransmitters involved in forming new connections between brain cells.

3 Make a gesture

There are also more leisurely ways to engage your body during learning, as the brain seems to find it easier to learn abstract concepts if they can be related to simple physical sensations. As a result, various experiments have shown that acting out an idea with relevant hand gestures can improve later recall, whether you are studying the new vocabulary of a foreign language or memorising the rules of physics.

It may sound dubious, but even simple eye movements might help. Andrew Parker and Neil Dagnall at Manchester Metropolitan University, UK, have found that subjects were better able to remember a list of words they had just studied if they repeatedly looked from left to right and back for 30 seconds straight after reading the list – perhaps because it boosts the transfer of information between the two brain hemispheres. It's worth noting, however, that this only seems to benefit right-handers. Perhaps the brains of left-handed and ambidextrous people already engage in a higher level of cross-talk, and the eye-wiggling only distracts them.

4 Engage your nose

Often, it's not just facts that we would like to remember, but whole events from our past as we reminisce about the good ol' days. Such nostalgia is not just an indulgence – it has been linked to a raft of benefits, such as helping us to combat lone-liness and feelings of angst. If you have trouble immersing yourself in your past, you could borrow a trick from Andy Warhol. He used to keep a well-organised library of perfumes, each associated with a specific period of his life. Sniffing each bottle reportedly brought back a flood of memories from that time – giving him useful reminders whenever he wanted to reminisce. Warhol's approach finds support in a spate of recent studies showing that odours tend to trigger particularly emo-tional memories, such as the excitement of a birthday party; they are also very effective at bringing back memories from our childhood. Some have even suggested that you could boost your performance in a test by sniffing the same scent during your revision and on the day of the exam.

5 Oil the cogs

Everyone's memory fades with age, but your diet could help you to keep your faculties for longer. You would do well to avoid high-sugar fast foods, for instance, which seem to encourage the build-up of the protein plaques characteristic of Alzheimer's disease.

In contrast, diets full of flavonoids, found in blueberries and strawberries, and omega-3 fatty acids, found in oily fish and olive oil, seem to stave off cognitive decline by a good few years – perhaps because the antioxidants protect brain cells from an early death.

6 Learn to forget

Sometimes we are haunted by unwanted memories: a moment of embarrassment perhaps, or a painful break-up. Banishing such recollections from our thoughts is difficult, but there may be ways of stopping fresh memories of painful events from being consolidated into long-term storage in the first place. For example, Emily Holmes at the University of Oxford asked subjects to watch a disturbing video, before asking them to engage in various activities. Those playing the video game Tetris subsequently experienced fewer flashbacks to unpleasant scenes in the film than those taking a general knowledge quiz, perhaps because the game occupied the mental resources usually involved in cementing memories. Playing relaxing music to yourself after an event you would rather forget also seems to help, possibly because it takes the sting out of the negative feelings that normally cause these events to stick in our mind.

3
Intelligence

Intelligence has always been tricky to pin down but that hasn't stopped generations of scientists trying to work out what it is, how to measure it, and why some people have more of it than others.

Quantifying intelligence: IQ

IQ is the best known of all the intelligence measures. It was developed in 1904 when the French Ministry of Education commissioned psychologist Alfred Binet to find a practical way to identify children who would fail elementary school without special help. The result was a short questionnaire with tasks such as naming an everyday object and identifying the heavier of two items. Performance on these tests, Binet believed, would indicate whether a child's learning was "retarded" relative to their peers. It worked well and led to a wave of intelligence-testing programmes on both sides of the Atlantic to screen: military recruits for trainability, college applicants for academic potential and job applicants for employability.

FIGURE 3.1 Alfred Binet devised the first IQ test

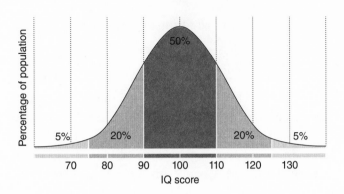

FIGURE 3.2 Average IQ score distribution by population

Their use remains controversial, partly because scores are easily affected by cultural or gender differences. The gold standard is the orally administered, one-on-one IQ test, which requires little or no reading and writing. Examples include the Stanford-Binet and Wechsler tests, which take between 30 and 90 minutes and combine scores from areas such as comprehension, vocabulary and reasoning to give an overall IQ. These tests are used to diagnose, treat or counsel children and adults who need personal or academic assistance and are governed by detailed ethical standards and strict criteria to avoid cultural bias. They must also be regularly updated. Done properly, IQ tests are the most technically sophisticated of all psychological tests and are still considered a useful way to measure variations in intelligence.

It's worth remembering that intelligence tests are calibrated so that, at a given age or group, the average IQ score is 100 and 90 per cent of individuals score between IQ 75 and 125. So an IQ score is always *relative* to the population a person is being compared to.

The g-factor – more than an IQ score

A century ago, British psychologist Charles Spearman observed that individuals who do well on one mental test tend to do well on all of them, no matter how different the tests' aims, format or content. So, for example, your performance on a test of verbal ability predicts your score on one of mathematical aptitude, and vice versa.

Spearman reasoned that all tests must therefore tap into some deeper, general ability and he invented a statistical method called **factor analysis** to extract this common factor from the web of positive correlations among tests. This showed that tests mostly measure the very same thing, which he labelled the general factor of intelligence or **g factor**. In essence, g equates to an individual's ability to deal with cognitive complexity.

Spearman's discovery lay neglected in the US until the 1970s, when psychologist Arthur Jensen began systematically testing competing ideas about g. Might g be a mere artefact of factor analysis? No, it lines up with diverse features of the brain, from relative size to processing speed. Might g be a cultural artefact, just reflecting the way people think in western societies? No, in all human groups – and in other species too – most cognitive variation comes from variation in g.

Jensen's analyses transformed the study of intelligence, but while the existence of g is now generally accepted, it is still difficult to pin down. Like gravity, we cannot observe intelligence directly, so must understand it from its effects. At the behavioural level, g operates as an indivisible force – a proficiency at mentally manipulating information, which undergirds learning, reasoning, and spotting and solving problems in any domain. At the physiological level, differences in g probably reflect differences in the brain's overall efficiency or integrity. The genetic roots of g are even more dispersed, probably emerging from the joint actions of hundreds if not thousands of genes, themselves responding to different environments.

Higher g is a useful tool: it is especially handy when life tasks are complex, as they often are in school and work. It is also broadly protective of health and well-being, being associated with lower rates of health-damaging behaviour, chronic illness, post-traumatic stress disorder, Alzheimer's and premature death.

It has little connection with emotional well-being or happiness, however. Neither does it correlate with conscientiousness, which is a big factor in whether someone fulfils their intellectual potential.

All kinds of clever

Consider the engineer's superior spatial intelligence and the lawyer's command of words and you have to wonder whether there are different types of intelligence. This question was debated ferociously during the early decades of the 20th century. Charles Spearman, on one side, defended the omnipotence of his general factor of intelligence, g. On the other, psychologist Louis Thurstone argued for seven "primary abilities", including verbal comprehension (in which females excel) and spatial visualisation (in which males excel). Thurstone eventually conceded that all his primary abilities were suffused with the same g factor, while Spearman came to accept that there are multiple subsidiary abilities in addition to g on which individuals differ.

This one-plus-many resolution was not widely accepted until 1993, however. It was then that American psychologist John B. Carroll published his "three stratum theory" based on a monumental reanalysis of all factor analysis studies of intelligence (*see* diagram, below). At the top is a single universal ability, g. Below this indivisible g are eight broad abilities, all

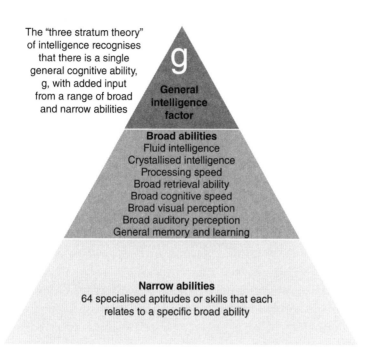

The "three stratum theory" of intelligence recognises that there is a single general cognitive ability, g, with added input from a range of broad and narrow abilities

g
General intelligence factor

Broad abilities
Fluid intelligence
Crystallised intelligence
Processing speed
Broad retrieval ability
Broad cognitive speed
Broad visual perception
Broad auditory perception
General memory and learning

Narrow abilities
64 specialised aptitudes or skills that each relates to a specific broad ability

FIGURE 3.3 The "three stratum theory" of intelligence

composed mostly of g but each also containing a different "additive" that boosts performance in some broad domain such as visual perception or processing speed. These in turn contribute to dozens of narrower abilities, each a complex composite of g, plus additives from the second level, together with life experiences and specialised aptitudes such as spatial scanning.

This structure makes sense of the many differences in ability between individuals without contradicting the dominance of g. For example, an excellent engineer might have exceptional visuospatial perception together with training to develop specialist abilities, but above all a high standing on the g factor. The one–plus–many idea also exposes the implausibility of multiple-intelligence theories eagerly adopted by educators in the 1980s, which claimed that by tailoring lessons to suit the individual's specific strength – visual, tactile or whatever – all children can be highly intelligent in some way.

What does a smart brain look like?

At Einstein's autopsy in 1955, his brain was something of a disappointment: it turned out to be a tad smaller than the average Joe's. Indeed, later studies have suggested a minimal link between brain size and intelligence. It seems brain quality rather than quantity is key.

One important factor seems to be how well our neurons can talk to each other. Martijn van den Heuvel, a neuroscientist at Utrecht University Medical Center in the Netherlands, found that smarter brains seem to have more efficient networks – in other words, it takes fewer steps to relay a message between different regions of the brain. That could explain about a third of the variation in a population's IQ, he says.

Another key factor is the insulating fatty sheath encasing neuron fibres, which affects the speed of electrical signals. Paul Thompson at the University of California, Los Angeles, has found a correlation between IQ and the quality of the sheaths.

And while people with the highest IQ do tend to have increased volume in a network of regions (*see* below), including key language areas, it may not be the size of your brain's processing hubs that counts, but how efficiently information flows between them.

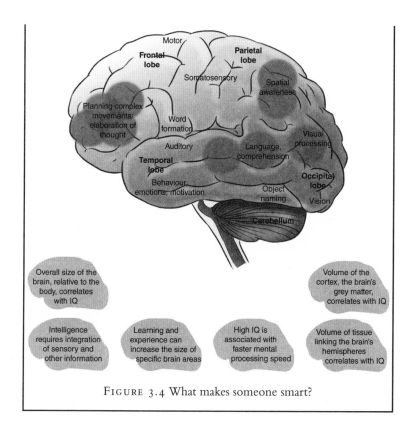

Motor

Frontal lobe

Parietal lobe

Somatosensory

Spatial awareness

Planning complex movements/ elaboration of thought

Word formation

Auditory

Temporal lobe

Visual processing

Language, comprehension

Behaviour, emotions, motivation

Object naming

Occipital lobe

Vision

Cerebellum

Overall size of the brain, relative to the body, correlates with IQ

Volume of the cortex, the brain's grey matter, correlates with IQ

Intelligence requires integration of sensory and other information

Learning and experience can increase the size of specific brain areas

High IQ is associated with faster mental processing speed

Volume of tissue linking the brain's hemispheres correlates with IQ

FIGURE 3.4 What makes someone smart?

Are humans getting more intelligent or more stupid?

In Denmark, every man is liable for military service at the age of 18. Nowadays, only a few thousand get conscripted but all have to be assessed, and that includes doing an IQ test. Until recently, the same one had been in use since the 1950s. "We actually have the same test being administered to 25,000 to 30,000 young men every year," says Thomas Teasdale, a psychologist at the University of Copenhagen.

"The results are surprising. Over this time, there has been a dramatic increase in the average IQ of Danish men. So much so that what would have been an average score in the 1950s is now low enough to disqualify a person from military service," Teasdale says.

The same phenomenon has been observed in many other countries. For at least a century, each generation has been measurably brighter than the last. But this cheerful chapter in social history seems to be drawing to a close. In Denmark, the most rapid rises in IQ, of about 3 points per decade, occurred from the 1950s to the 1980s. Scores peaked in 1998 and have actually declined by 1.5 points since then. Something similar seems to be happening in a few other developed countries, too, including the UK and Australia.

So why have IQ scores been increasing around the world? And, more importantly, why does this rise now seem to be coming to an end? The most controversial explanation is that rising IQ scores could have been hiding a decline in our genetic potential. Could this possibly be right? Do we face a future of gradually declining intellectual wattage?

There's no question that intelligence – as measured by IQ tests, at least – has risen dramatically since the tests were first formalised a century ago. In the US, average IQ rose by 3 points per decade from 1932 to 1978, much as in Denmark. In post-war Japan, it shot up by an astonishing 7.7 points per decade, and two decades later it started climbing at a similar rate in South Korea. Everywhere psychologists have looked, they have seen the same thing.

The Flynn effect

This steady rise in test scores has come to be known as the "Flynn effect" after James Flynn of the University of Otago in New Zealand, who was one of the first to document the trend.

Much has been written about why this has been happening. There may be a cultural element, with the rise of television, computers and mobile devices making us better at certain skills. The biggest IQ increases involve visuospatial skills. Increasing familiarity with test formats may also play a role.

The general view, though, is that poor health and poor environments once held people back, and still do in many countries. Wherever conditions start to improve, though, the Flynn effect kicks in. With improved nutrition, better education and more stimulating childhoods, many people around the world really have become smarter.

We have, after all, changed in other ways: each generation has been taller than the previous one, probably because nutrition has improved. So although height is thought to have an even larger genetic component than intelligence – taller parents tend to have taller children – the environment matters too.

If better nutrition and education have led to rising IQs, the gains should be especially large at the lower end of the range, among the children of those with the fewest advantages in their lives. Sure enough, that's what testers usually see. In Denmark, for example, test scores of the brightest individuals have hardly budged – the score needed for an individual to place in the top 10 per cent of the population is still about what it was in the 1950s. Only the lowest scores saw a rise.

If social improvements are behind the Flynn effect, then as factors like education and improved nutrition become common within a country their intelligence-boosting effects should taper off, country by country. "I've been predicting for some time that we should see signs of some of them running out," says Flynn. And those signs are indeed appearing. It seems we are seeing the beginning of the end of the Flynn effect in developed countries.

Intelligence: nature or nurture?

It's the eternal question that has a simple answer: both. Each of us is the embodiment of our genes and the environment, which work together from conception to death.

To understand how these two forces interact to generate differences in intelligence, behavioural geneticists compare twins, adoptees and other family members. The most compelling research comes from identical twins adopted into different homes – same genes, different environment – and unrelated people who were adopted into the same home – different genes, same environment. These and other studies show that IQ similarity most closely lines up with genetic similarity.

More intriguingly, the studies also reveal that the heritability of intelligence – the percentage of its variation in a particular population that can be attributed to its variation in genes – steadily increases with age. Heritability is less than 30 per cent before children start school, rising to 80 per cent among western adults. In fact, by adolescence, separated identical twins answer IQ tests almost as if they were the same person and adoptees in the same household as if they were strangers.

The surprising conclusion is that most family environments are equally effective for nurturing intelligence – the IQ of an adult will be the same almost regardless of where he or she grew up, unless the environment is particularly inhumane.

Why do the shared environments' power to modify IQ variation wane and genetic influences increase as children gain independence? Studies on the nature of nurture offer a clue. All children enter the world as active shapers of their own environment. Parents and teachers experience this as their charges frustrate attempts to be shaped in particular ways. And increasing independence gives young people ever more opportunities to choose the cognitive complexity of the environments they seek out. The genetically brighter an individual, the more cognitively demanding the tasks and situations they tend to choose, and the more opportunities they have to reinforce their cognitive abilities.

IQ decline

But IQ scores are not just levelling out but appear to be declining.

The first evidence of a small decline, in Norway, was reported in 2004 (*see* graph, below). Since then a series of studies have found similar declines in other highly developed countries, including Australia, Denmark, the UK, Sweden, the

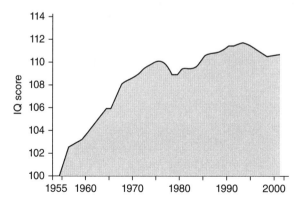

The rise of average IQ scores of military conscripts in Norway has slowed and started to reverse. Similar patterns are seen in a few other countries

FIGURE 3.5 The pattern of IQ scores among military conscripts in Norway

Netherlands and Finland. Should we be worried? Not according to Flynn and Teasdale. The evidence remains sparse and sometimes contradictory, and could just be due to chance.

"Even if they are not down to chance, such small declines could be attributable to minor changes in social conditions, such as falling income or poorer education, which can easily be reversed," says Flynn. But these are invented hypotheses for a very small phenomenon, he points out. "You'd want to be pretty certain that the phenomenon was actual before you scratch around too hard for causes."

There is a more disquieting possibility, though. A few researchers think that the Flynn effect has masked an underlying decline in the genetic basis of intelligence. In other words, although more people have been developing closer to their full potential, that potential has been declining.

Most demographers agree that in the past 150 years, in Western countries, the most highly educated people have been having fewer children than is normal in the general population. The notion that less educated people are outbreeding others is far from new, as is the inference that we are evolving to be less intelligent. It's even the theme of a 2006 film, *Idiocracy*. "This is a claim that has been made for over a century now, and always with the most horrific prediction of what might happen if we don't stop it," says Bill Tucker, a historian of psychology at Rutgers University in Camden, New Jersey. This idea led to the extensive eugenics programme in the US, with its forced sterilisations, which in turn helped inspire the "purity" policies of Nazi Germany.

This unpleasant history, though, doesn't mean there is no genetic decline, some argue. Richard Lynn of the University of Ulster, UK, a psychologist whose work has often been controversial, has tried to calculate the rate of decline in our genetic potential using measured IQ values around the world in 1950 and 2000. His answer: a bit less than 1 IQ point, worldwide, between 1950 and 2000. If the trend continues, there would be another 1.3 point fall by 2050. Even if he is right – and it's a big if – that is a tiny change compared with the Flynn effect. Would small declines like this even matter?

Yes, argues Michael Woodley, a psychologist at Free University of Brussels (VUB) in Belgium. This kind of evolution would shift the bell curve of intelligence, he claims, and a small shift can lead to a big drop in the number of high scorers. For example, if mean IQ fell from 100 points to 97, it would almost halve the number of people who score above IQ 135.

Bigger-brained babies

Would this really matter? People who score highly in IQ tests are not always the most successful in life. In any case, with so many confounding factors it is far from clear whether the "evolving to be stupid" effect is real. For example, it has been suggested that caesarean sections allow more bigger-brained babies to survive than in the past.

A definitive way to settle this issue would be to look at whether gene variants associated with higher IQs are becoming less common. The trouble with this idea is that so far, despite huge effort, we have failed to find any specific gene variant linked to significantly higher IQs in healthy individuals.

Yet Woodley thinks his team has found clear evidence of a decline in our genetic potential – and he claims it is happening much faster than Lynn's calculation suggests. Instead of relying on fertility estimates, Woodley looked at a simple measure: reaction time. Quick-witted people, it turns out, are exactly that: smarter people tend to have quicker reaction times, probably because they process information more quickly.

Back in the 1880s, the polymath Francis Galton measured the reaction times of several hundred people of diverse social classes in London. A few years ago, Irwin Silverman of Bowling Green State University in Ohio noticed that the reaction times Galton recorded – an average of about 185 milliseconds between seeing a signal and pushing a button – were quite a bit quicker than the average of more than 250 milliseconds in modern tests, which began in the 1940s.

Woodley's team reanalysed Silverman's data, factoring in the known link between reaction time and intelligence. When they did this, they found that reaction times had indeed slowed over the century, by an amount corresponding to the loss of one full IQ point per decade, or more than 13 points since the Victorian era.

Critics have been quick to attack Woodley's analysis, arguing that Galton may not have measured reaction times in the same way as later investigators. If Galton's apparatus had a button with a shorter range of motion, for instance, then he would have measured shorter reaction times. What's more, Silverman points out that there is no obvious downward trend in the post-1940 data, as there should be if Woodley is right.

Variation in IQ

In a detailed response published in June 2014, Woodley maintains that today's brains remain slower even after accounting for all these other explanations. But even if he's right about reaction times, the correlation between IQ and reaction time is not an especially strong one: reaction time explains only about 10 per cent of the variation in IQ.

"Probably every generation moans about the new generation being less intelligent, and every upper crust moans about the lower classes out-breeding them," says Kevin Mitchell, a neurogeneticist at Trinity College Dublin in Ireland. "The basic premise is that IQ levels are dropping. And I don't see any evidence for that, which is why I find the whole debate a bit odd."

The coming decades should provide a definitive answer. If what we are seeing in countries like Denmark is merely the end of the Flynn effect, IQ scores should stabilise in developed countries. If Woodley and his colleagues are right, we should see a continuing decline.

Even if we are evolving to be more stupid, it is far from clear whether we need to worry about it. Flynn thinks the problem may just take care of itself, as societal improvements such as better healthcare and more promising employment options bring down fertility rates in every stratum of society.

In the longer term, there may be an even more fundamental threat to our intelligence. We humans mutate fast − each of us has 50 to 100 new mutations not present in our parents, of which a handful are likely to be harmful, says Michael Lynch, an evolutionary geneticist at Indiana University in Bloomington. In the past, harmful mutations were removed as fast as they appeared, because people unlucky enough to inherit lots of them tended to die young, before they had children. Now, things are different. Foetal mortality, for example, has declined by 99 per cent in England since the 1500s, Lynch says.

This means that populations in developed countries are accumulating harmful mutations.

Over tens of generations, Lynch has calculated, this will lead to a large drop in genetic fitness. With so many genes contributing to brain function, such a decline might well drag down our brainpower, too. The only way to stop that might be to tinker with our genomes. Given our ignorance about the genetic basis of intelligence, and the ethical complexities, that is a long way off.

Coming back to the short-term, though, there is an obvious option for those concerned about intelligence levels. "If you're worried about it, the answer is what the answer has always been," says Mitchell. "Education. If you want to make people smarter, educate them better. That won't make everybody equal, but it will lift all boats."

What if intelligence is a dead end?

Our intelligence, the very trait we like to think makes us the pinnacle of evolution, could be our undoing. What if, evolutionarily speaking, being stupid is better?

Humans have evolved a unique form of intelligence, with cognitive complexity unseen in other species. This has been the secret behind our agricultural, scientific and technological progress. It has let us dominate a planet and understand vast amounts about the universe. But it has also brought us to the brink of catastrophe: climate change looms and a mass extinction is already under way, yet there is little sign of a concerted effort to change our ways.

Our troubles could be compounded by the fact that human genetic diversity is abysmally low. "One small group of chimpanzees has more genetic diversity than the entire human species," says Michael Graziano of Princeton University. It's not unthinkable that a global disaster could wipe us out.

For this, we have an awkward double-act to blame. says philosopher Thomas Metzinger of the University of Mainz, Germany. He argues that we have reached this point because our intellectual prowess must still work alongside hardwired primitive traits. "It is cognitive complexity, but without compassion and flexibility in our motivational structure," says Metzinger.

In other words, we are still motivated by some rather basic instincts, such as greed and jealousy, and not by a desire for global solidarity, empathy or rationality. And it's unclear whether we will evolve the necessary social skills in time to thwart planetary disaster.

Another part of the problem is that our intelligence comes with so-called cognitive biases. For instance, psychologists have shown that humans pay less attention to future risk compared with present risk, something that makes us routinely take decisions that are good in the short term but disastrous in the long term. This may be behind our inability to fully fathom the risks of climate change, for example.

Humans also have what philosophers call **existence bias**, which influences our view of the value of life – it's better to exist than not. But what if our intelligence were to develop in a way that meant we lost such biases?

Metzinger, muses that perhaps super-intelligent aliens may have already achieved that. With a balanced outlook no longer weighted to the short term and a clear-eyed view of suffering, such a life form could decide that life is just not worth it. "They may have come to the conclusion that it's better to terminate their own existence," says Metzinger.

Could that explain why we haven't yet made contact with an alien intelligence? "Possibly," he says.

4
Emotions

Of all the outputs of our brains, emotions are perhaps the most mysterious. We know them when we feel them and seek them out in our lives and choice of entertainment. Yet while we all know how they feel and are remarkably good at recognising them in others, the whys and hows of emotion raise some important questions about what it means to be human.

First, the biggest question of all: do emotions feel the same to every human on the planet and, if so, why might that be?

Emotions: written all over our faces

In 1868, while working on his latest book on evolution, Charles Darwin liked to show visitors to his house a series of ghoulish photographs of people's faces.

The pictures, taken by French physiologist Guillaume-Benjamin Duchenne, showed people whose facial muscles were being zapped by electric shocks, contorting them into strange and often eerie arrangements. Darwin was fascinated by how a twitch of the mouth here or furrow of the brow there could conjure up the impression of an emotion – fear, say, or surprise. He wanted to know whether his guests perceived the same emotions in the pictures. They usually did.

FIGURE 4.1 An expressive Serena Williams

Darwin came to the conclusion that emotional facial expressions were universal: people all over the world made the same ones, and could easily and automatically recognise them in others. He didn't claim to know what the expressions were for – he thought they were probably "not of the least use" – but he did suggest that they were innate and rooted in our shared ancestry with other animals. He presented this argument to the world in his book *The Expression of the Emotions in Man and Animals*, published in 1872.

Darwin wasn't the first scientist to investigate the meaning of facial expressions, but his enormous influence brought them to wider attention and initiated a debate that has ebbed and flowed ever since. Are emotional expressions universal and innate, or do they vary from culture to culture?

By the late 20th century, the pendulum appeared to have swung decisively in favour of Darwin's view. But today the debate is alive and well again and has implications for understanding what emotions actually are, and what they tell us about human nature.

The modern orthodoxy on emotional expressions was largely secured by a classic experiment that took place a century after Darwin was spooking his visitors.

In the late 1960s, a team of psychologists led by Paul Ekman of the University of California in San Francisco travelled to Brazil, Japan, Borneo and New Guinea and showed people photographs of six stereotypical emotional expressions: happiness, fear, anger, surprise, disgust/contempt and sadness. They did the same experiment at home in the US.

Ekman's team found that everybody they tested, regardless of culture, recognised the same six basic emotions: joy, sadness, anger, fear, surprise and disgust – even people in Borneo and New Guinea who'd had no prior contact with the outside world. The research, published in *Science* in 1969, strongly supported the idea that expressions of emotions are common to all people, regardless of culture, because they have a common evolutionary origin.

Emotional hardwiring

Since then, dozens of studies have seen similar effects. In addition to the basic six (or seven, with contempt and disgust sometimes treated separately), the model has been extended to include pride, indicated by a tall posture and puffed-up chest, and shame, with a downturned head and bent posture. This all supports the view that emotional expressions are hardwired into the human brain.

Other support comes from studies of people who were born blind and have therefore never seen an emotional expression. For example, psychologist David Matsumoto at San Francisco State University analysed the facial expressions of judo contestants at the 2004 Olympics and Paralympics, including athletes who were blind from birth or later became blind. He found that all three groups produced the same faces when they won a bout. The expressions included so-called Duchenne smiles – big, beaming smiles involving the eyes as well as the mouth – which are considered to be authentic expressions of happiness.

If emotional expressions are universally made – and read – it begs the question of how they evolved in the first place. One idea is that the characteristic facial movements that accompany emotions have physiological functions. When people make a classic fear expression, for example, they increase their field of vision, make more rapid eye movements, and open their airways – all of which would allow them to better monitor danger and consequently respond.

Plausible functions have also been proposed for other emotions. The scrunched-up expression of disgust, for example, may serve to constrict airways to stop the entry of contaminants, while the cringing posture of shame may hide vulnerable parts of the face from attack. Not all emotional expressions have obvious functions, however. The original biological purpose of happy smiles, angry scowls or sad frowns have so far eluded psychologists.

The origin of facial expressions goes beyond simple physical responses, however. Humans are social animals who need to

According to one theory, all humans make and recognise the same basic emotional expressions, Each probably has a biological function

Emotional expression		Proposed physiological function	Proposed communicative function
Happiness		Unknown	Communicates a lack of threat
Sadness		Unknown	Tears handicap vision to signal appeasement and elicit sympathy
Anger		Unknown	Alerts of impending threat and communicates dominance
Fear		Widened eyes increase visual field and speed up eye movements	Alerts to possible threat and appeases potential aggressors
Surprise		Widened eyes increase visual field to see unexpected stimulus	Unknown
Disgust		Constricted orifices reduce inhalation of possible contaminant	Warns about aversive foods or distasteful ideas and behaviours
Pride		Testosterone and lung capacity increase to prepare for confrontation	Communicates heightened social status
Shame		Reduces/hides bodily targets from potential attack	Communicates lessened social status and desire to appease

FIGURE 4.2 Emotional expressions

communicate, and facial expressions are a very powerful way of doing so. Being able to transmit and receive emotional states would have been advantageous to our ancestors. For example, displaying fear and reading it on another face helps both of you to respond to danger.

In this scenario, emotional expressions started out as something evolutionary biologists call a **cue** – they revealed information about an inner state or behaviour but they didn't evolve for this signalling purpose, much as chewing is a reliable signal that somebody is eating. Then, over time, they evolved into signals for expressly conveying information. Expressions became more exaggerated and distinctive to make it easier to communicate non-verbally.

This process may explain why it is hard to discern a function for some expressions: the original purpose has been lost in translation. Another possibility is that some emotional expressions only ever served a signalling function. Pride and shame, which are particularly social emotions, are likely candidates. The expressions resemble the dominance and submission postures of other social primates, suggesting they are signals of status inherited from distant ancestors.

To complicate matters further, some researchers question the idea that emotional expressions are universal at all. They point out that the study methods used by Ekman and others, where volunteers are given a list of emotions and asked to pick the one that best matches the facial expression they see. Sceptics argue that if subjects know that they are trying to identify happiness, sadness, anger and so on, then that is probably what they will see. When people are asked to come up with their own words, they find it much harder to hit the target. In one experiment, removing the list reduced accuracy from more than 80 per cent to about 50.

Instead, some researchers argue that rather than being biologically based, our emotional expressions are culturally learned symbols – a form of "body language" that we learn to communicate emotions to others. And, like spoken languages, the expressions share commonalities but also vary from culture to culture.

Experiments have shown that people's recognition of an emotion depends heavily on context. In the real world, faces are rarely seen in isolation. Posture, voices, other faces and the wider context are also available for inspection, and they influence how expressions are perceived. For instance, a scowl – usually associated with anger – can be perceived as disgust if the person is holding a dirty object, or fear if it's paired with a description of danger. The disgust face can even be seen as pride if attached to a body with arms raised in triumph. Similarly, viewing the same expression labelled alternately with the words "anger", "surprise" and "fear" changes how people perceive it.

It also turns out that, contrary to Ekman's classic paper, emotional expressions are not culturally invariant. "There are differences," says Rachael Jack at the University of Glasgow, UK. In a 2012 study she used a graphics package to generate thousands of expressions by randomly combining facial muscle positions.

Her team generated 4,800 faces and showed them to 15 European and 15 Chinese volunteers. Their task was to categorise the faces as expressing one of the six basic emotions or to say "don't know", with no predetermined correct answers. The Europeans (who were shown European-looking faces) reliably sorted the expressions into clusters representing the six basic emotions, with high levels of agreement between them. But the Chinese (who saw east Asian faces) produced much more overlapping categorisations, and disagreed much more.

"I wouldn't disagree with people who suggest that there is a biological origin to certain facial expressions," Jack says, "but people have had culture for about 80,000 years." These once-hardwired signals have been extensively reshaped by cultural evolution for use in social communication, she says, allowing regional variations to arise.

So if emotional expressions turn out to be less universal than Ekman and others claim, what's the alternative? One idea is that when we observe emotions in others, the categories we use are culturally constructed, learned and context-dependent. There is also evidence that immigrants gradually adapt their emotions to the norms of their new home. So it could be that we speak more or less the same language, but adopt the local dialect to communicate best with the people around us.

Can we feel emotions without words to describe them?

The idea that the labels we give our feelings might influence how we feel them is hotly debated. Some evolutionary psychologists believe that long before they learned to speak, our cave-dwelling ancestors would have felt the tell-tale physiological fear response – their hearts would hammer in their chest and their palms prickle with sweat if they watched a sabre-toothed tiger slope past. In this scheme of things, feelings came first, their names much later, as people learned to communicate. If so, "disgust" feels the same, whether you live in New York or Timbuktu.

Look closely at the world's languages, however, and the idea of universal emotions takes a tumble. If disgust is a single primal emotion, why do Germans distinguish between two types of it – *ekel* (disgust that makes the gorge rise or stomach churn) and *abscheu* (usually translated as revulsion)? And that is nothing compared with the 15 kinds of fear that the Pintupi of Western Australia speak of.

Some cultures identify feelings which have no obvious equivalent in English, like the Japanese *amae*, the comforting feeling of being unconditionally loved and cared for, or the Dutch *gezelligheid*, which describes both the physical state of being in a homely place surrounded by good friends and the emotional state of feeling "held" and comforted.

Equally, languages may lack words for emotions which English-speakers take for granted: the Machiguenga people of Peru have no term which precisely captures the meaning of "worry", for instance. Could the fact that they don't have a word to convey this emotion mean they don't – or can't – feel it either?

Name that feeling

Science is now being brought to bear on such questions. Brain imaging studies, for example, show a strong link between language and emotions: when the parts of the brain linked to emotion are aroused, so are those parts associated with semantics and language.

Some of these studies have shown what many of us know instinctively, that putting a name to a feeling can soothe us, bringing coherence to internal turbulence. Other cognitive scientists have gone further, suggesting that words play an even deeper role in constructing our emotional lives, not only helping us manage feelings, but actually bringing them into being in the first place.

There is some evidence that this might be the case. People with semantic dementia, a neurodegenerative disorder in which people lose the meaning of words, have shown that when words for certain emotions get lost, it becomes more difficult to recognise them in others. Given a pile of photographs showing emotional faces, healthy people will sort them into the big six emotions. Those with semantic dementia tend to make three piles: one for unpleasant emotions, another for pleasant ones and a third for neutral.

Without the words to understand emotions, then, we might not even be able to register faces as expressing different feelings at all. Similar processes may be at work in recognising our own feelings. When we learn a word for an emotion, it may act as a lightning rod, attracting all kinds of inchoate sensations and vague inklings. Once we learn to link that word to a particular network of sensations, our brains find it easier to seek out experiences which are consistent with it and filter out those which aren't.

It works both ways. Some feelings go unnamed and so stay unnoticed. As far as the conscious mind is concerned, they are unfelt. It might even make sense to say that when a language lacks the name for an emotion, the feeling can fade into the background, unformed, even lost.

If true, this finding has important therapeutic consequences. Recently, Jordi Quoidbach at Pompeu Fabra University in Barcelona, Spain, and his colleagues found that "emodiversity" – experiencing an abundance and wide range of emotions – is strongly correlated with long-term emotional and physical health.

So if you want to bring some variety into your emotional life, try familiarising yourself with *greng jai*: a Thai word to describe being reluctant to accept help from someone because of the bother it would cause them; or *iktsuarpok*: the fidgety feeling when expecting visitors, and you might just notice yourself experiencing new feelings as a result in your daily life too. Just beware of *basorexia* – the sudden urge to kiss someone.

Why do we cry?

Look closely at crying, and you will see just how strange it is. For one thing, it encompasses two very different processes: vocal wailing and tearing.

Human babies excel at the former, and for good reason – bawling is a very effective way of getting attention from caregivers. For their first few weeks, babies don't even shed tears, because their tear glands are still developing. But as they grow, crying becomes less vocal and more tearful.

This could be an evolutionary adaptation, suggests Ad Vingerhoets at Tilburg University in the Netherlands. Wailing advertises vulnerability to everyone around, including predators, so once a child can move around, it is wiser to use the more covert signal of tears.

Another puzzle is that we cry throughout our lives. Intriguing changes in crying behaviour seem to reflect its changing functions as we age. Around adolescence, we begin to cry less over physical pain and more over emotional pain. Many people also start to exhibit "moral crying", in reaction to acts of bravery, self-sacrifice and altruism. Why we do this is still a mystery.

Also mysterious is why, as we age, we increasingly shed tears over things that are positive. Robert Provine at the University of Maryland, Baltimore County, has a suggestion. "Given that emotional tearing is recently evolved, it's a very crude estimate of emotional expression," he says.

Another theory is that so-called tears of joy do not actually reflect happiness at all; events such as weddings and holidays are often bittersweet because they remind us of the passage of time and mortality. This may be why children usually do not cry out of happiness: they don't yet make the associations with sacrifice, loss and impermanence.

Then there's the question of why some people cry more than others. In a recent review of research, Vingerhoets reported that people who score highly on measures of neuroticism and those who are empathic cry the most. The former use tears manipulatively – as do narcissists, psychopaths and tantrum-throwing toddlers. Sociopaths are thought most likely to cry fake or "crocodile" tears.

And, although boys and girls cry frequently until puberty, in Western cultures women cry at least twice as often as men. Men are culturally conditioned to restrain their tears, but there may be more to it than that. Studies in animals suggest the hormone testosterone might have a tear-suppressing effect.

Three steps to emotional mastery

Mastering your emotions is important not just for psychological well-being, but also for success in many areas of life. Happily, psychologists have identified three skills that can help us all become more emotionally adept, and reap the benefits.

Emotions evolved to help animals react quickly in life-or-death situations. But, says Mark Pagel, an evolutionary biologist at the University of Reading, UK, human emotions are complicated by our social lives. "We have jealousy, sympathy, a sense of injustice, and guilt. It's these social emotions which really mark us out as a species." They are also what make our emotional lives so complicated.

Some people are clearly better at coping with this complexity than others. This might help explain why the idea of **emotional intelligence** (EQ) was so eagerly received in 1995, following the publication of psychologist Daniel Goleman's book *Emotional Intelligence: Why it can matter more than IQ*. An international bestseller, it launched an industry peddling tests to select emotionally intelligent candidates for management positions and careers such as medicine. But for all the hype and the money spent, there has been a sense of disappointment that it hasn't delivered on its early promise.

Part of the confusion comes from the fact that while notionally EQ suggests a fixed measure akin to IQ, even its proponents promise that employees, students, indeed anyone, can learn to boost their score. Many psychologists now prefer the term "emotional competence" because it signifies an ability that can be honed.

Many also think of this ability as a sort of language: one that all humans share. Just as learning a language entails recognising words, understanding how to use them and controlling a conversation, so mastering the language of emotions requires three key skills: perception, understanding and regulation of emotions.

Perception

Perception is the bedrock on which the two other skills rest. Perceiving emotions is not as straightforward as it might sound. Traditional tests of emotional intelligence probe this skill using pictures of faces, but expressions of emotion extend beyond the face to gestures and movements, plus tone of voice and other sounds. Aural and visual cues can interact; for example, one study found that the way people interpret laughter and crying sounds is altered by the facial expressions accompanying them.

A static picture isn't even a good representation of the way our faces express emotion. "The human face is equipped with a large number of independent muscles, each of which can be combined and activated at different levels of intensity over time," says Rachael Jack at the University of Glasgow, UK. Her studies using computer-generated faces that randomly combine facial expressions, such as lip curls and raised eyebrows, suggest that each emotion has an associated sequence of facial movements, which she calls "action units", unfolding a bit like the letters of a word. Action units strung together in specific patterns create "sentences" that communicate a more complex social message.

So, how can you improve your emotion recognition skills? Some studies suggest that tailored training might help. In one study, people who were trained to look for the appropriate cues in the face, voice and body improved their emotional recognition skills compared to controls. Others have been looking at whether musical training can help. One study has found that adult musicians are better than non-musicians at judging the emotion in someone's tone of voice. Brain imaging studies suggest that this reflects more than simply a general sensitivity to basic aspects of sound, but that music training may modulate brain responses known to be more specifically associated with emotions.

Understanding

The next step in emotional mastery is to understand how emotions are used. The catch is that emotional signals vary a lot from one person to another. "Not everyone smiles when they're happy, or scowls when they're angry," says Lisa Feldman Barrett, at Northeastern University. Barrett has found tremendous variability in brain activity, both between people and in the same individual, in response to different types of threat.

This suggests that there is no "essence" of particular emotions. A person who is fluent at understanding emotions – both from the outside world or their own bodies – can take emotional signals and make sense of them.

"This is a language that has to be taught," says Marc Brackett, director of the Yale Center for Emotional Intelligence. A decade ago, in an attempt to do just that, he helped create a programme called RULER, now used in some 10,000 US schools. It teaches children and young adults to interpret physiological changes in their bodies linked to emotions, label them, and learn strategies to regulate their emotions. "It's remarkable work that has a tremendous impact on kids' competence," says Barrett. "When you can take a physical change in your body and understand it as an emotion, you learn to make meaning out of that change."

Regulation

The final skill in emotional mastery is the ability to regulate your feelings. Again, this isn't something we are born with, and as we develop some of us learn ineffective strategies for doing it, such as avoiding emotionally charged situations or trying to shut down our emotions completely. Research shows that people who address emotional situations directly rather than avoiding them have higher levels of well-being and are better able to cope with stress.

There are ways to improve your regulation skills. One approach that psychologists favour is **reappraisal** – trying to put yourself in someone else's shoes so as to be more objective, and change your emotional response accordingly. When a team led by Ute Hülsheger at Maastricht University in the Netherlands taught this strategy to hairdressers, waiters and taxi drivers, they found that it resulted in more tips.

But rethinking your emotions from scratch requires a lot of effort. Another promising approach is **mindfulness** – observing the coming and going of your emotions without action or judgement. In a separate study, Hülsheger randomly picked members of a group of 64 employees to receive mindfulness training, and monitored them all over ten days. Those who got the training reported more job satisfaction and less emotional exhaustion. "The idea is that when you just see emotions as they are, as thoughts and sensations, you gain a sense of perspective and the 'hot' aspect of the emotion dissolves," she says.

Everyone knows that mastering a language takes time and practice. Some people are naturals. Others struggle to communicate effectively. But when it comes to the language of emotions, making the effort to improve is surely worth it, because the proponents of emotional intelligence were right about one thing – being emotionally fluent really does bring benefits.

Five emotions you never knew you had

Officially, there may only be six basic human emotions, but in recent years, though, a few more obscure emotions have been tentatively added to the list, each of which may serve equally important functions in modern life. Here we explore five candidates that could be promoted alongside the Big Six.

1 **Elevation:** the feeling of being uplifted, inspired and positive. It has been documented in Japan, India, the US and the Palestinian territories. Its proposed function is to bring people together, feeling warm and positive, through the release of the hormone oxytocin.

2 **Interest:** muscles in the forehead and eyes contract, this is the expression we wear when trying to understand something new. It may motivate learning and help prevent information overload in unfamiliar situations.

3 **Gratitude:** a feeling that motivates us to acknowledge or repay kindness or favours and may be useful in maintaining long-term relationships based on give and take.

4 **Pride:** throw head backwards, arms out wide, and take up as much space as possible – pride is unmistakable. It comes in two forms: "hubristic pride", the boastful kind that establishes your status; and "authentic pride", which follows hard work and achievement. In body language terms, though, they look the same.

5 **Confusion:** easily spotted with a furrowed brow, narrowed eyes, maybe a bitten lip. Confusion tells us that we are lacking information or that our mental model needs to be updated. Its job is to motivate change and maybe call on others to help.

5
Sensation and perception

We are only just beginning to understand how our brain combines the wealth of sensory information that bombards it into one seamless experience. How is all that information processed, and how does the brain cope when it receives conflicting information? And could new technology provide a way to access a whole new world of sensory information?

A sensory primer

We navigate the world without giving it a second thought. Yet at every instant, our senses are being bombarded by stimuli, which our brains are turning into perceptions of organised, meaningful objects and events.

Take the example of riding a motorbike. It is an experience packed with sensory experiences: the visual sensation of motion, the green tint of trees, the blue of the sky, the pressure of the wind, the sound and vibration of the engine. These elementary sensations are then put together into a more sophisticated understanding of what you are doing. Your perception is not a series of disconnected sensations, but a coherent whole: that you are riding a bike.

How does the brain perform this astonishing feat? The first stage of the process is to collect sensory information, something that requires minimal effort on our part, but which relies on a huge amount of brain activity, most of which happens beneath the radar of consciousness.

Despite being almost trivially easy, this skill is so sophisticated it has proved impossible to design computers that do the job anywhere near as well. Airport security officers inspecting luggage scans or doctors studying an x-ray for a tumour are good examples of skills where the sensory system allows us to make very subtle and complex decisions, which no supercomputer has yet been able to match. The same can be said of the countless sensory processes that go on in every brain, all the time. We are all equipped with a highly sensitive set of sensory kit, and we don't even need to know how it works.

What we are very aware of, however, is that we have five basic senses: taste, smell, vision, hearing and touch – a system first identified by the ancient Greeks. But scientifically, five isn't the most plausible number. Instead of dividing the senses by how we experience them, we could break it down according to the way the brain deals with them. We could break it down to four senses because the brain has four kinds of receptor cells for the senses: light (sight), chemical compounds (taste and smell), mechanical force (hearing, touch and balance) and tissue damage (pain). Or to seven senses because there are seven different nerve pathways carrying sensory signals through the brain, including the olfactory nerve and the optic nerve.

Then, once the information has flowed through the brain and reached its destination, there are six different areas of the cortex that process the information, each dedicated to a certain kind of processing, from visual to auditory, somatosensory (body sensations), olfactory (smell), gustatory (taste) and vestibular (balance).

So if we are talking about senses in the brain, it makes a lot of sense to talk about six, not five. It's possible that these areas

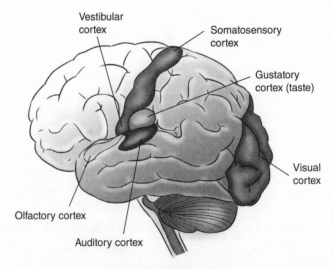

FIGURE 5.1 Cortical sensory areas

could be divided even further – the somatosensory cortex covers both touch and pain as well as the movement and position of limbs in space, for example, which feel very different and so could count as separate "senses".

Another difficulty in defining the senses comes from the fact they often meld together, making it difficult to discern how much they each contribute to a specific perception.

If you were to stand on one leg, for example, you might think that is mostly dealt with by vestibular system, which deals with balance. But if you ask people to close their eyes while standing on one leg, it cuts down the amount of time they can stay balanced to a quarter of what they could manage with their eyes open. In fact, we use visual input for almost everything. When you're walking along high ridges on top of mountains, the reason you may experience vertigo is that the detail of the surroundings doesn't change much in your visual field. In effect you're walking along ridges with your eyes closed.

Similarly, the flavour we get from food is not only driven by taste receptors in the mouth but also olfactory receptors in the nose and back of mouth, and the way the food feels and looks – the gloopiness of yogurt and fizz of lemonade, for example. That's why food manufacturers are just as interested in getting the right texture and colour for their food as they are the taste. At least part of the taste experience happens before we even put food in our mouth.

Combining the senses

How the brain combines these senses into one seamless experience is only partially understood. We do know that some integration happens in the brain stem. There are direct reflex circuits that connect vision and balance, for example, so that when you nod your head your eyes automatically move in their sockets to maintain clear vision. More recently we have discovered that the brain has direct connections between the six different sensory regions in the cortex, potentially allowing different areas of the brain to share information.

A third way is that other areas of the cortex receive projections from several senses and merge this information together so coherently that you wouldn't be able to tell the difference between a visual or auditory signal.

This is a seriously huge task: the brain is receiving information at an enormous rate: there are more than 200 million sensory receptor cells, fewer than 3 million sensory fibres and 16 billion cortical cells, all putting in their ten cents' worth at any one time and the brain only has a certain amount of storage and energy available to deal with this information.

Even while sitting at home watching television, making sense of the world is an enormous challenge. Your brain consumes 20 per cent of the energy budget of the body, but the blood can only supply the brain with enough energy to keep 3 per cent of its neurons highly active at any one time. It's a bit like turning on the lights only in the room of your house that you're in. The brain shuffles around energy to the parts of the brain that you're using at the time. Three per cent may not sound much, but it is about 500 million brain cells.

So the brain has a problem – the information coming from the outside world vastly outstrips the amount of storage and energy available. The brain deals with this by throwing away most of the information it receives.

It has a couple of tricks up its sleeve to achieve this without glitches appearing in the system. The first trick is that it does the throwing away unconsciously. You don't know what you don't know, so it doesn't feel like anything is missing. The brain

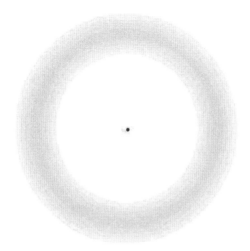

FIGURE 5.2 Troxler's fading

is pretty good at knowing which bits it can throw away at any given instant. Most of the time, for example, we are completely unaware of the weight of our limbs and the feeling of clothes against our skin. Our sensory systems are throwing away this information because it's not changing in the short term, and the brain banks on the fact that it is therefore probably not important.

The same is true for other senses. A well-known optical illusion called **Troxler's fading** shows this in action. If you stare at the fixed point in the image the images on the periphery disappear after a short time. This happens because the sensory system assumes that the edges of a scene are less important than where you are focusing your eyes.

Having to throw away so much sensory information means that there is never enough to figure out what is going on around you without having to resort to guesses. The good news is that the brain is a very good guesser – by combining incoming sensory information with knowledge based on prior experience it is able to make sense of less-than-perfect information.

Scientists study this process using a statistical method known as **Bayesian inference** – a method that works on the basis of probability (*see* Chapter One). The brain draws inferences completely automatically and with no conscious awareness, so we don't really appreciate that so much of what we experience is actually based on highly educated guesses.

Ultimately, we see, hear and feel partly on the basis of what we expect to see, hear and feel. Different people generally agree about what they see and hear, but because everything is based on unconscious interpretation, which can vary between people, there can sometimes be legitimate disagreements. That's why opposing players and supporters can disagree so wildly about the same refereeing decision. The brain has thrown away the details and the guesses it makes to fill in the gaps might be very different indeed.

The sound of dots moving

It sounds like a Zen conundrum: what is the sound of dots moving? Most of us can't answer that question, but for people with synaesthesia, hearing such sights is the most natural thing in the world.

Melissa Saenz, a neuroscientist at Caltech in Pasadena, was tipped off when a visitor looked at her screensaver, which made no noise, and said: "Does anyone else hear that?" She quizzed him and found that his experience had the hallmarks of synaesthesia: a trigger through one sense was giving rise to a sensory experience in another. Some people perceive letters, numbers, words and smells to have innate colours, while others can taste music or imagine time to have a fixed special form.

It isn't clear what proportion of the population experiences synaesthesia – most synaesthetes have always experienced the world this way and are surprised that others do not. Since everyone's senses are intricately linked in the brain it is possible that it is more common than we yet realise.

Perception of time: How your brain creates now

What is "now"? We tend to think of it as this current instant, a moment with no duration. But if now were timeless, we wouldn't experience a succession of nows as time passing. Neither would we be able to perceive things like motion. We couldn't operate in the world if the present had no duration. So how long is it?

FIGURE 5.3 Split-second timing relies on the brain's ability to predict which perceptual stimuli belong together

That sounds like a metaphysical question, but neuroscientists and psychologists have an answer. In recent years, they have amassed evidence indicating that now lasts on average between 2 and 3 seconds. This is the now you are aware of – the window within which your brain fuses what you are experiencing into a "psychological present". It is surprisingly long. But that's just the beginning of the weirdness. There is also evidence that the now you experience is made up of a jumble of mini subconscious nows and that your brain is choosy about what events it admits into your nows. Different parts of the brain measure now in different ways. What's more, the window of perceived now can expand in some circumstances and contract in others.

Pinning down the slippery concept of now could tell us something about the bigger picture of how the brain tracks time and the way that we experience the world and understand what causes what. "Your sense of nowness underpins your entire conscious experience," says Marc Wittmann at the Institute for Frontier Areas of Psychology and Mental Health in Freiburg, Germany.

The brain has many ways of experiencing different timescales. It has long been known that the brain contains structures that use cycles of light and dark to set its daily circadian clock. How it tracks the passing of seconds and minutes is much less well understood. At this level, there are two broad types of timing mechanism, an **implicit** and an **explicit** one. The explicit one relates to how we judge duration – something we're surprisingly good at. The implicit mechanism is the timing of "now" – it is how the brain defines a psychological moment and so structures our conscious experience.

Our implicit sense of time is itself made up of two seemingly incompatible aspects: the fact that we exist permanently in the present yet experience time flowing from the past towards the future. Wittmann believes that the brain does this by constructing a hierarchy of nows, each of which forms the building blocks of the next, until the property of flow emerges.

If Wittmann is correct, to understand the now that we experience we first need to understand its subconscious component, the "functional moment", which operates on the timescale at which a person can distinguish one event from another. This varies for different senses. The auditory system, for example, can distinguish two sounds just 2 milliseconds apart, whereas the visual system requires tens of milliseconds. Detecting the order of stimuli takes even longer. Two events must be at least 50 milliseconds apart before you can tell which came first.

The brain must somehow reconcile these different detection thresholds to make sense of the world. Its task is made more difficult by the fact that light and sound travel through air at different speeds and can reach our sensory apparatus at different times, even if they were emitted by the same object at the same time.

To some extent, the brain is able to compensate for this. When you watch a badly dubbed movie, the brain predicts that the audio and visual streams should occur simultaneously and – as long as the lag between them doesn't exceed about 200 milliseconds – that's what you perceive. After a while you stop noticing that the lip movements and voices of the actors are out of sync. Virginie van Wassenhove at the French medical research agency's Cognitive Neuroimaging Unit in Gif-sur-Yvette and her colleagues looked at how this works in the brain, exposing people to sequences of beeps and flashes, both occurring once per second, but 200 milliseconds out of sync, while recording electrical activity in the brain.

They found two distinct brainwaves, one in the auditory cortex and another in the visual cortex, both oscillating at a frequency of 1 hertz – once per second. At first the two oscillations were out of phase, and the volunteers experienced the

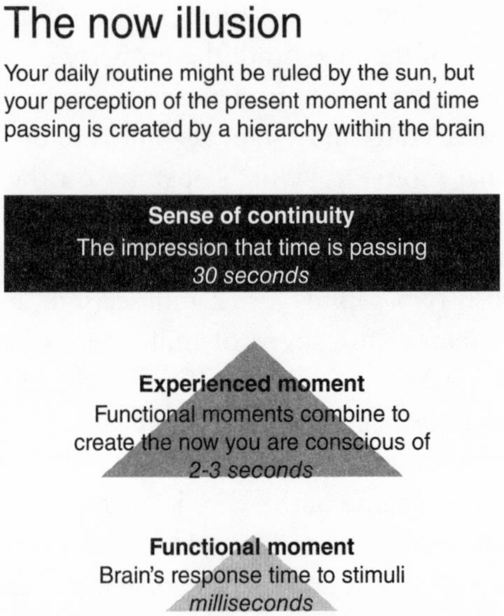

The now illusion

Your daily routine might be ruled by the sun, but your perception of the present moment and time passing is created by a hierarchy within the brain

Sense of continuity
The impression that time is passing
30 seconds

Experienced moment
Functional moments combine to create the now you are conscious of
2-3 seconds

Functional moment
Brain's response time to stimuli
milliseconds

FIGURE 5.4 The now illusion

light and sound as out of sync. But as people reported that they started to perceive the beeps and flashes as being simultaneous, the auditory oscillation became aligned with the visual one. The brain seems to physically adjust signals to synchronise events if it decides that they should belong together.

This is the first time that a biological basis has been found for implicit timing. It also suggests that, even at the subconscious level, the brain is choosing what it allows into a moment. However, this functional moment is not the now of which we are conscious. That comes at the next level of Wittmann's hierarchy, with the "experienced moment". So what do we know about this?

It is this now that seems to last between two and three seconds. A neat demonstration of that was provided recently by David Melcher at the University of Trento, Italy, and his colleagues. They presented volunteers with short movie clips in which segments lasting from milliseconds to several seconds had been subdivided into small chunks that were then shuffled randomly. If the shuffling occurred within a segment of up to 2.5 seconds, people could still follow the story as if they hadn't noticed the switches. But the volunteers became confused if the shuffled window was longer than this. The researchers suggest that this window is the "subjective present", and exists to allow us to consciously perceive complex sequences of events.

Melcher thinks the two-to-three second window provides a sort of bridging mechanism to compensate for the fact that our brains are always working on outdated information. Right now, your brain is processing stimuli that impinged on your senses hundreds of milliseconds ago, but if you were to react with that lag you wouldn't function effectively in the real world.

Consciously or not, Hollywood movie editors take account of our experienced moment. In the cutting room, they rarely create shots that last less than two or three seconds, unless the director is aiming to create a sense of chaotic or confusing movement. "Three seconds is long enough to understand what's going on, but not so long that you have to rely too heavily on memory to maintain access to all the relevant information," says Melcher. "It's the sweet spot."

Wittmann acknowledges that it is not clear how a group of subconscious functional moments are combined to create the conscious experienced moment.

Creating flow

However the present moments we experience arise, they are combined to give us a sense of continuity or "mental presence", the final now in Wittmann's hierarchy. This operates over a timespan of about 30 seconds and gives a sense of continuity. According to his model, the glue that holds the experienced moments together to create an impression of time flowing is working memory – the ability to retain and use a limited amount of information for a short time. Mental presence is what underpins the sense that it is you who is experiencing events. "It is the now of 'I', of your narrative self," Wittmann says.

The implications of this new view of nowness are potentially mind-boggling. Take, for example, the debate over free will. In the 1980s, US physiologist Benjamin Libet found that people reported deciding to flick their wrist about 500 milliseconds after he had detected activity in their brains that preceded each wrist-flick. His now-controversial conclusion was that we have less conscious control over our actions than we think. But, given what we know about implicit timing, it is possible that what he actually detected was an artefact of the brain's insensitivity to order at very small time scales. At 500 milliseconds, says Wittmann, "we are definitely within margins of temporal resolution where you cannot distinguish which event came first".

Then there's the issue of the stretchiness of now. There is plenty of anecdotal evidence that time can seem to expand or contract depending on what's happening around us – for example, that events seem to unfold in slow motion during car accidents. Such expansion has been reproduced in the lab, when people are presented with a succession of stimuli of equal length yet report that an oddball event in the series seems to have a longer duration. What's more, Melcher has preliminary findings showing that when people perceive an event to have lasted longer than it actually did, they also take in more detail about it, describing it more accurately. In his opinion, this shows that temporal stretchiness reflects real changes in sensory processing, which in turn may have conferred an evolutionary advantage. By ratcheting up the brain's processing rate at critical moments and easing back when the environment becomes predictable and calm again, we conserve precious cognitive resources.

Such changes in sensory processing would be subconscious, but might we be able to take control of our perception of now? Regular meditators often claim that they live more fully or intensely in the present than most people. In experiments Wittmann asked 38 people who meditate and 38 who do not to look at an ambiguous line drawing of a cube, known as a Necker cube, and press a button each time their perspective of it reversed. The reversal time in this kind of task is considered a good estimate of the length of the psychological present. By this measure, people in both groups perceived now to last about four seconds, seeming to confound the claims of some meditators. However, when Wittmann asked participants to try to hold a given perspective for as long as possible, the meditators managed eight seconds on average, compared with six seconds for the others.

Meditators tend to score highly in tests of attention and working memory capacity, says Wittmann. "If you are more aware of what is happening around you, you not only experience more in the present moment, you also have more memory content." And that in turn affects your sense of the passing of time. "Meditators perceive time to pass more slowly than non-meditators, both in the present and retrospectively," he says.

This suggests that with a bit of effort we are all capable of manipulating our perception of now. If meditation extends your now, then as well as expanding your mind it could also expand your life. So, grab hold of your consciousness and revel in the moment for longer. There's no time like the present.

Interview: The future of perception: pick a sense and bolt it on

Our window on the world is limited by what a human body can perceive with its handful of senses. Neuroscientist **David Eagleman** *is trying to change all that. He has invented a vibrating jacket that the brain can use as a new kind of sensory information. In theory, he says, you can programme it to add any sense you like.*

You have described the brain as "locked in a vault of silence and darkness", so how does it create such a rich reality for us to experience?

That's one of the great mysteries of neuroscience: how do electrochemical messages in your brain get turned into your subjective experience of the world? What we know is that the brain is good at extracting patterns from our environment and assigning meaning to them. I'm interested in how we can plug alternative patterns into the brain and experience additional aspects of reality.

What new realities could we perceive?

We only pick up on a small fraction of signals that are going on in the world: those for which we have developed specialised sensors. There are many other signals out there – X-rays and gamma rays among others – but we're completely blind to them. No matter how hard we try, we'll never see that part of the spectrum naturally.

But the brain is really flexible about what it can incorporate into its reality. It receives information in the form of electrochemical signals from our eyes, our nose, our skin, and works out meaning from them. Crucially, it doesn't care where these signals are coming from; it just figures out how to use them.

I think of the brain as a general-purpose computer. Our senses are just plug-and-play devices that we have inherited through evolution. And if that's the case, we should be able to interface any data stream into the brain and it will figure out how to deal with it.

So how do you plan to patch a new data source into the brain?

We're experimenting with what we call the versatile extrasensory transducer, or VEST. It's a wearable device covered with vibratory motors. When you wear the VEST, at first it just feels like strange patterns of vibrations on your torso, but the brain is really good at unlocking sensory patterns and figuring out what the input means.

How does a vibrating jacket allow us to experience a different reality?

Well, for example, we are trialling it with deaf participants at the moment. We capture sound from their environment and translate it into different patterns of vibrations. After a week or so our volunteers are able to figure out what's being said using the vibrations alone. They can understand the auditory world through their skin.

But the brain is specialised to hear different frequencies. Can it really figure out speech from vibrations on the skin?

It seems crazy to hear via a moving pattern of touch on the skin, but ultimately this is just translated into electrochemical signals coursing round the brain – which is all that regular hearing ever is. Traditionally, these signals come via the auditory nerve, but here they come via nerves in the skin. We already know that the brain can figure out meaning from arbitrary signals. For example, when a blind person passes their fingers over Braille the meaning is directly evident to them. And when you read words in a *New Scientist* article, you don't have to think about the details of the squiggles – the meaning simply flows off the page. In the same way, we're demonstrating that a deaf person can extract the meaning of words coming to them by vibratory patterns.

The vibrating jacket has so many potential applications. What else have you in mind?

Yes, it's hard to decide which possibilities to test first. We are playing with feeding the jacket a real-time sentiment analysis on Twitter, as filtered by a hashtag. Let's say you're a presidential candidate giving a speech. You could wear the jacket and feel how the Twittersphere is reacting as you're going along. One of our other experiments involves working with pilots and feeding them cockpit data or drone information via the VEST.

That's intriguing. Do the pilots end up feeling like they've become the drone, say?

Yes, it's like extending your skin to the plane or drone, so you feel the pitch, yaw and roll. It's a new perceptual experience that we believe will allow someone to pilot better. We're thinking about astronauts too. They spend a lot of time looking at hundreds of monitors, so wouldn't it be great if they could directly feel the state of the space station and know when things were changing or shifting?

Our whole lives are spent looking at little screens. In my view it's better to experience the data rather than simply look at it.

Are there limits to how many extra senses you might acquire using wearables?

You mean, could you have a Twitter jacket and stock-market jeans? I don't see why not. We don't know if there are any limits to how many different things you could sense. My intuition is that the limits are distant. We've got an enormous amount of real estate in the brain. If you lose one sense, the area of the brain responsible for it gets taken over by other senses. The brain is great at redistributing space for what's needed and there's plenty of room to share real estate without noticing any diminished effects elsewhere.

Does the VEST system have limitations?

The thing that defines what is possible is what we can build sensors for. If we have a good sensor, it's trivial to convert the information it captures into vibrations in the jacket.

If you could choose only one extra sense to have, what would it be?

That's an interesting question. Right now, everything about our society is engineered around the senses that we currently have. If I were suddenly able to have ultrasonic hearing, I would hear animal calls that no one else could hear. As a nature lover that would be amazing, but I don't know if it would be lonely in that extrasensory space if no other human joins me there.

I'd also like to explore whether the VEST can allow us to better connect with other people. Perhaps if my wife and I both wore a VEST, and used it to somehow experience each other's emotions, that might bring us to a new level of closeness. Or perhaps it would be detrimental [laughs] – we just don't know until we try.

David Eagleman is a neuroscientist at Stanford University, California, and author of *Brain: The story of you* and other books.

6
Consciousness

There are a lot of hard problems in the world, but only one of them gets to call itself "the hard problem". That is the problem of consciousness – how a kilogram or so of nerve cells conjures up the seamless kaleidoscope of sensations, thoughts, memories and emotions that occupy every waking moment.

The intractability of this problem prompted British psychologist Stuart Sutherland's notorious 1989 observation: "Consciousness is a fascinating but elusive phenomenon… Nothing worth reading has been written on it."

The hard problem remains unsolved. Yet neuroscientists have still made incredible progress understanding consciousness, from the reasons it exists to the problems we have when it doesn't work properly.

Is consciousness still fascinating? Yes. Elusive? Absolutely. But Sutherland's final point no longer stands. Read on…

Your brain on consciousness

What does consciousness "look like" in the brain? Discovering the physical or neural correlates of consciousness is one area in which we have made great progress – what consciousness in the brain "looks like", you might say. One way to investigate this question is to see what changes when consciousness is reduced or absent, as happens when people are in a vegetative state, with no sign of awareness.

Brain scans show that such people usually have damage to the **thalamus**, a relay centre located smack-bang in the middle of the brain (*see* diagram below). Another common finding

Brain scanning reveals that three areas of the brain play a pivotal role

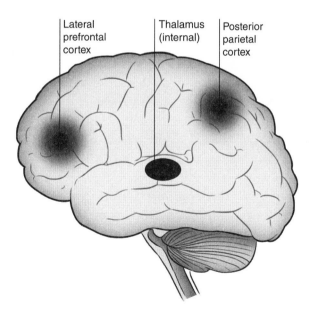

| Lateral prefrontal cortex | Thalamus (internal) | Posterior parietal cortex |

FIGURE 6.1 Seats of consciousness

is damage to the connections between the thalamus and the **prefrontal cortex**, a region at the front of the brain, generally responsible for high-level complex thought.

The prefrontal cortex has also been implicated using another technique – scanning the brain while people lose consciousness under general anaesthesia. As awareness fades, a discrete set of regions are deactivated, with the **lateral prefrontal cortex** the most notable absentee.

Those kinds of investigations have been invaluable for narrowing down the search for the parts of the brain involved in us being awake and aware, but they still don't tell us what happens in the brain when we see the colour red, for example.

Seeing red

Simply getting someone to lie in a brain scanner while they stare at something red won't work, because we know that there is lots of unconscious brain activity caused by visual stimuli – indeed, any sensory stimuli. How can we get round this problem?

One solution is to use stimuli that are just at the threshold of awareness, so they are only sometimes perceived – playing a faint burst of noise, for instance, or flashing a word on a screen almost too briefly to be noticed. If the person does not consciously notice the word flashing up, the only part of the brain that is activated is that which is directly connected to the sense organs concerned, in this case the visual cortex. But if the subject becomes aware of the words or sounds, another set of areas kicks into action. These are the lateral prefrontal cortex and the **posterior parietal cortex**, another region heavily involved in complex, high-level thought, this time at the top of the brain, to the rear.

Satisfyingly, while many animals have a thalamus, the two cortical brain areas implicated in consciousness are nothing like as large and well developed in other species as they are in humans. This fits with the common intuition that, while there may be a spectrum of consciousness across the animal kingdom, there is something very special about our own form of it.

In humans the three brain areas implicated in consciousness – the thalamus, lateral prefrontal cortex and posterior parietal cortex – share a distinctive feature: they have more connections to each other, and to elsewhere in the brain, than any other region. With such dense connections, these three regions are best placed to receive, combine and analyse information from the rest of the brain. Many neuroscientists suspect that it is this drawing together of information that is a hallmark of consciousness. When I talk to a friend in the pub, for instance,

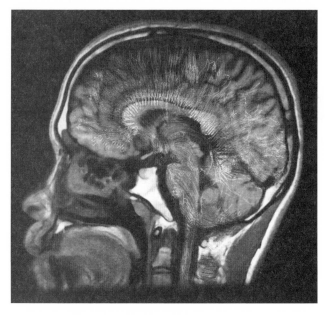

FIGURE 6.2 In theory we could calculate how conscious anything is, be it human, rat or computer

I don't experience him as a series of disjointed features, but as a unified whole, combining his appearance with the sound of his voice, knowledge of his name, favourite beer and so on – all amalgamated into a single person-object.

How does the brain knit together all these disparate strands of information from a variety of brain locations? The leading hypothesis is that the relevant neurons start firing in synchrony many times a second, a phenomenon we can see as brainwaves on an electroencephalogram (EEG), whereby electrodes are placed on the scalp. The signature of consciousness seems to be an ultrafast form of these brainwaves originating in the thalamus and spreading across the cortex.

One of the most prominent attempts to turn this experimental data into a theory of consciousness is known as the "global neuronal workspace" model. This suggests that input from our eyes, ears and so on is first processed unconsciously, primarily in sensory brain regions. It emerges into our conscious awareness only if it ignites activity in the prefrontal and parietal cortices, with these regions connecting through ultrafast brainwaves.

This model links consciousness with difficult tasks, which often require a drawing together of multiple strands of knowledge. This view fits nicely with the fact that there is high activity in our lateral prefrontal and posterior parietal cortices when we carry out new or complex tasks, while activity in these areas dips when we do repetitive tasks on autopilot, like driving a familiar route.

The main rival to global workspace as a theory of consciousness is a mathematical model called the "information integration theory", which says consciousness is simply combining data together so that it is more than the sum of its parts. This idea is said to explain why my experience of meeting a friend in the pub, with all senses and knowledge about him wrapped together, feels so much more than the raw sensory information that makes it up. But the model could be applied equally well to the internet as to a human: its creators make the audacious claim that we should be able to calculate how conscious any particular information-processing network is – be it in the brain of a human, rat or computer. All we need to know is the network's structure, in particular how many nodes it contains and how they are connected together.

Unfortunately, the maths involves so many fiendish calculations, which grow exponentially as the number of nodes increases, that our most advanced supercomputers could not perform them in a realistic time frame for even a simple nematode worm with about 300 neurons. The sums may well be simplified in future, however, to make them more practical.

This mathematical theory (see The phi factor, below) may seem very different from global neuronal workspace – it ignores the brain's anatomy, for a start – yet, encouragingly, both models say consciousness is about combining information, and both focus on the most densely connected parts of the information-processing network. This common ground reflects the significant progress the field is making.

We may not yet have solved the so-called hard problem of consciousness – how a bunch of neurons can generate the experience of seeing the colour red. Yet in many ways, worrying about the hard problem is just another version of dualism – seeing consciousness as something that is so mysterious it cannot be explained by studying the brain scientifically.

Every time in history we thought there had to be some supernatural cause for a mysterious phenomenon – such as mental illness or even the rising of bread dough – we eventually found the scientific explanation. It seems plausible that if we continue to chip away at the "easy problems" we will eventually find there is no hard problem left at all.

The phi factor

Perhaps the best way to understand the mathematical integration theory is to consider the difference between the brain and a digital camera. Although the screen seems to show a complete image to our eyes, the camera just treats the image as a collection of separate pixels, which work completely independently from one another; it never combines the information to find links or patterns. For this reason, it has very low "integration" or as the theory's creator, Giulio Tononi of the University of Wisconsin–Madison, calls it, "phi".

The brain, on the other hand, is constantly drawing links between every bit of information that hits our senses, so has high "phi".

"Now I could go back to neurobiology with this tentative theory: any seat of consciousness must have a high level of phi, and other systems must not," says Tononi.

Some accepted anatomical findings gel with this tentative theory. For instance, we know that the cerebral cortex is crucial for conscious experience – any damage to the brain here will have an effect on your mental life. Conversely, the cerebellum is not necessary for conscious awareness, which was something of a puzzle given that it contains more than twice as many neurons as the cerebral cortex.

When Tononi analysed the two regions using his theory, it all made sense: the cerebral cortex may have fewer neurons, but the cells are very well connected to one another. They can hold large amounts of information and also integrate it to generate a single coherent picture – the level of phi is very high. The cerebellum is more like the digital camera: it may contain more neurons than the cerebral cortex, but there are fewer interconnections and so no coherent picture – the level of phi is low, in other words.

"I've been studying consciousness for 25 years, and Giulio's theory is the most promising," says Christof Koch at the California Institute of Technology in Pasadena. "It's unlikely to be the final word, but it goes in the right direction – it makes predictions. It moves consciousness away from the realm of speculative metaphysics."

Tononi's theory can also explain what happens when we fall asleep or are given an anaesthetic – through experiments he has shown that the level of phi in the cerebral cortex drops as our consciousness fades away.

Is there such a thing as an unconscious mind?

Humans are rather proud of their powers of conscious thought – and rightly so. But there is one aspect of our cognitive prowess that rarely gets the credit it deserves: a silent thinking partner that whirrs away in the background, without bothering to inform our conscious mind of the details: the unconscious.

While Freud's view of the unconscious as a repository of our repressed desires is no longer accepted, there is a huge amount of evidence that our brains do a huge amount of processing under the radar of consciousness.

An early example of this came from an experiment done in the 1980s by Benjamin Libet at the University of California, San Francisco (*see* diagram, below). People were told to wait a little while and then press a button whenever they liked, and to note the exact time they decided to act on an ultra-precise clock. They also had electrodes placed on their scalp to measure electrical activity in their brain.

This set-up revealed that neuronal activity preceded people's conscious decision to press the button by nearly half a second. More recently, a similar experiment placed people in an fMRI scanner instead of hooking them up to electrodes. This found stirrings in the brain's prefrontal cortex up to ten seconds before someone became aware of having made a decision.

These results are sometimes interpreted as disproving the existence of free will. On the other hand, it could mean that we do have free will but that it is our unconscious mind that is, in fact, in charge, not the conscious one. Neuroscientist John-Dylan Haynes at the Max Planck Institute in Leipzig, Germany, who led the brain scanning study, warns against jumping to this conclusion. "I wouldn't interpret these early [brain] signals as an 'unconscious decision'," he says. "I would think of it more like an unconscious bias of a later decision."

The fact that unconscious processing is, by definition, something that we are unaware of is quite difficult to study in the lab. One technique is to use a method called "masking", in which an

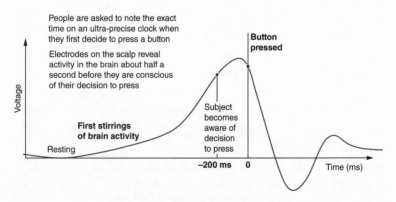

This experiment seems to challenge the notion of free will

People are asked to note the exact time on an ultra-precise clock when they first decide to press a button

Electrodes on the scalp reveal activity in the brain about half a second before they are conscious of their decision to press

Voltage

Button pressed

First stirrings of brain activity

Subject becomes aware of decision to press

Resting

−200 ms   0

Time (ms)

FIGURE 6.3 Who's in charge? An experiment on free will

image is flashed in front of the eyes only to be quickly replaced with another before the first image can consciously register. In this way it has been demonstrated that information shown to the unconscious can spill over into conscious thoughts and decisions. For instance, people shown the masked word "salt" are then more likely to select a related word, like "pepper", from a list.

Asking people to choose words from a list might seem a rather artificial test, but such unconscious associations can spill over into life outside the research lab. One study, for instance, showed that people acted more competitively in a game if the pen and paper were taken out of a briefcase rather than a rucksack. Afterwards no one was aware of it affecting their behaviour.

If these ideas are disconcerting, there may be an upside, points out Ap Dijksterhuis of Radboud University Nijmegen in the Netherlands. Our ability to unconsciously process information may help us to make decisions.

In one study by Dijksterhuis, people were asked to choose an apartment by one of three methods. These were: making an instant decision, mulling over all the pros and cons for a few minutes, or thinking about an unrelated problem in order to distract them from consciously thinking about the apartments. People chose the objectively best apartment when they used the distraction method. Dijksterhuis thinks that this is because they were unconsciously mulling over the decision while their consciousness was elsewhere.

Some of these findings have recently been questioned as others have been unable to replicate them. Yet there is certainly growing attention paid to the powers of the unconscious. Dijksterhuis reckons that unconscious deliberation can also explain those "a-ha!" moments when the answer to a problem seems to come from nowhere, as well as times when a searched-for word comes to mind only after we stop trying.

Many neuroscientists agree that, whether or not the unconscious is capable of complex thought, it almost certainly does
more for us than we ever realise. "For all sorts of decisions we
are never aware of all the myriads of influencing factors," says
Haynes.

7
Ages and sexes of the brain

Your brain undergoes profound changes as you age. We now have a clear understanding of this process and can begin to answer questions such as why you can't remember much from your early years and what babies think about. Recent research also sheds light on another conundrum: Are male and female brains really different?

The five ages of the brain

Throughout life our brains undergo more changes than any other part of the body. These can be broadly divided into five stages, each profoundly affecting our abilities and behaviour. But we are not just passengers in this process. There are some ways to get the best out of our brains at every stage and pass the best possible organ on to the next.

In utero: Setting the stage

By the time we take our first breath, the brain is already more than eight months old. It starts to develop within four weeks of conception, when one of three layers of cells in the embryo rolls up to form the neural tube. A week later the top of this tube bends over, creating the basic structure of fore, mid and hindbrain.

From this point, brain growth and differentiation is controlled mainly by the genes. Even so, the key to getting the best out of your brain at this stage is to have the best prenatal environment possible. In the early weeks of development, that means having a mother who is stress-free, eats well and stays away from cigarettes, alcohol and other toxins. In fact, if you consider the size of the construction job at hand – 100 billion brain cells and several million support cells in four major lobes and tens of distinct regions – it is a truly staggering feat of evolutionary engineering.

One nutrient we know the brain needs early on is folic acid, which is crucial for closing the neural tube. Deficiencies can lead to defects like spina bifida, where part of the spine grows outside the body, and anencephaly, a fatal condition in which much of the brain fails to develop. Animal studies suggest that malnutrition – particularly a lack of protein – stunts the growth of neurons and connections, and that iron and zinc are needed for neurons to migrate from where they form to their final location. Long-chain polyunsaturated fatty acids are required for synapse growth and membrane function. Usually this is taken care of by a healthy diet but an inefficient placenta – caused by high blood pressure, stress or smoking – can hinder development. Too much of one nutrient can also be bad news. Poorly controlled diabetes can cause a potentially toxic excess of glucose in the developing brain.

Towards the end of the brain-building process, when the foetus becomes able to hear and remember, experience also begins to shape the brain.

Childhood: Soaking it up

In childhood, the brain is the most energetic and flexible that it will ever be. As we explore the world around us it continues to grow, making and breaking connections at breakneck speed. Learning can be detected even before we are born. At about 22 to 24 weeks of gestation, foetuses will respond to a noise or a touch but will ignore the same stimulus if it occurs repeatedly – a simple kind of memory called **habituation**. From around 32 weeks foetuses show **conditioning** – a more complex kind of memory in which an arbitrary stimulus can be learned as a signal that something will happen, like a sound signalling a poke. Foetal memories for particular pieces of music and the mother's voice, smell and language have all been shown to form some time after 30 weeks' gestation and to persist after birth.

Birth alters brain function surprisingly little. Although the touch-sensitive somatosensory cortex is active before birth, it's another two or three months before there is any other activity in the cortex, which ultimately governs such things as voluntary movement, reasoning and perception. The frontal lobes become active between six months and a year old, triggering the development of emotions, attachments, planning, working memory and attention. A sense of self develops as the parietal and frontal lobe circuits become more integrated, at around 18 months, and a sense of other people having their own minds at age three to four.

Life experiences in these early years help shape our emotional well-being, and neglect or harsh parenting may change the brain for good. Maternal rejection or trauma early in life, for example, may affect a person's emotional reactions to stressful events later on, potentially predisposing them to depression and anxiety disorders.

The good news for parents is that there is no reason for your child to stop playing and start working. Studies have shown that a nurturing environment and one-on-one playtime with games like peekaboo, building blocks, singing nursery rhymes and shape-sorting are all a child needs to increase IQ and foster a lifelong interest in learning.

By age six, the brain is 95 per cent of its adult weight and at its peak of energy consumption. Around now, children start to apply logic and trust and to understand their own thought processes. Their brains continue to grow and make and break connections as they experience the world until, after a peak in grey matter volume at 11 in girls and 14 in boys, puberty kicks in and the brain changes all over again.

Why don't we remember our early years?

"Childhood amnesia", as the phenomenon is known, is universal. Most people remember nothing from before the age of two or three, and memories from the next few years are sketchy at best.

This is puzzling, because in other ways children are phenomenal learners. In our first couple of years we pick up many complex, lifelong skills, like the ability to walk, talk and recognise people's faces. Yet memories of specific events in our childhood are lost to us in adult life. It's as if someone has torn the first few pages from our autobiography.

So what causes childhood amnesia? Part of the answer seems to come down to the way the brain develops. Two major structures are involved in the creation and storage of autobiographical memories: the prefrontal cortex and the hippocampus. The hippocampus is thought to be where details of an experience are cemented into long-term memory. One small area of the hippocampus, called the **dentate gyrus**, does not fully mature until age four or five. This area acts as a kind of bridge that allows signals from the surrounding structures to reach the rest of the hippocampus, so until the dentate gyrus is up to speed early experiences may never get locked into long-term storage.

Other factors seem to be the age at which a child develops a sense of self: the understanding that the entity "me" is different from "you". This ability emerges at around 18 to 24 months of age, about the age when it starts becoming possible to remember the past. Language also seems to play a part. In experiments, children's ability to remember their recent past was linked to the number of words they could speak and understand. One idea is that we need to have a word to describe a concept before we can lay down a memory of it.

Once all of these pieces of the puzzle are in place, a child is able to start constructing a narrative of their own lives. Parents play a big role in this – studies have shown that children whose parents talk to them in more detail about what is happening around them tend to form earlier memories than parents who don't do this. Overall, it looks as if language and self-perception go hand in hand, and both are necessary for autobiographical memory to flourish.

Why we can't access our earliest memories as adults is an open question. Some think that they were never laid down as detailed memories in the first place. Another possibility is that they are in there somewhere, just not easily accessible. Some think that our pre-language memories are stored as "snapshots" of sensory experiences. If that's the case, buried memories could be excavated – if we could only find the right cues.

Adolescence: Wired, and rewiring

Teenagers are selfish, reckless, irrational and irritable, but given the construction going on inside the adolescent brain is it any wonder? Psychologists used to explain the particularly unpleasant characteristics of adolescence as products of raging sex hormones, since children reach near-adult cerebral volumes well before puberty. More recently, though, imaging studies have revealed a gamut of structural changes in the teens and early 20s that go a long way towards explaining these tumultuous teenage years.

Jay Giedd at the National Institute of Mental Health in Bethesda, Maryland, and his colleagues have followed the progress of nearly 400 children, scanning many of them every two years as they grew up. They found that adolescence brings waves of grey-matter pruning, with teens losing about one per cent of their grey matter every year until their early 20s.

This cerebral pruning trims unused neural connections that were overproduced in the childhood growth spurt, starting with the more basic sensory and motor areas. These mature first, followed by regions involved in language and spatial orientation and lastly those involved in higher processing and executive functions.

Among the last to mature is the **dorsolateral prefrontal cortex** at the very front of the frontal lobe. This area is involved in control of impulses, judgement and decision-making, which might explain some of the less-than-stellar decisions made by your average teen. This area also acts to control and process emotional information sent from the **amygdala** – the fight or flight centre of gut reactions – which may account for the mercurial tempers of adolescents.

As grey matter is lost, though, the brain gains white matter (*see* graph, below). This fatty tissue surrounds neurons, helping to conduct electrical impulses faster and stabilise the neural connections that survived the pruning process.

These changes have both benefits and pitfalls. At this stage of life the brain is still childishly flexible, so we are still sponges for learning. On the other hand, the lack of impulse control may lead to risky behaviours such as drug and alcohol abuse, smoking and unprotected sex.

So while they carry with them the raw potential to sculpt their brains into lean, mean processing machines, whether they like it or not, their decision-making circuits are still forming – tender teen brains still need to be protected, if only from themselves.

While pruning of underused pathways means we lose grey matter (brain cells), an increase in white matter speeds electrical impulses and stabilises connections

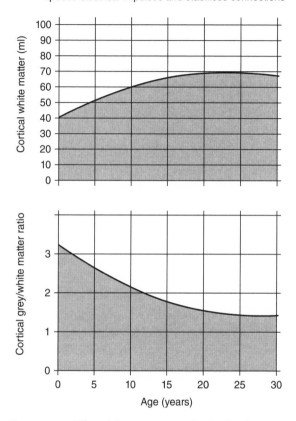

FIGURE 7.1 The adolescent stage of brain development

Adulthood: The slippery slope

So you're in your early 20s and your brain has finally reached adulthood. Enjoy it while it lasts. The peak of your brain's powers comes at around age 22 and lasts for just half a decade. From there it's downhill all the way.

This long, slow decline begins at about 27 and runs throughout adulthood, although different abilities decline at different rates. Curiously, the ones that start to go first – those involved with executive control, such as planning and task coordination – are the ones that took the longest to appear during your teens. These abilities are associated with the prefrontal and temporal cortices, which are still maturing well into your early 20s.

Episodic memory, which is involved in recalling events, also declines rapidly, while the brain's processing speed slows down and working memory is able to store less information.

So just how fast is the decline? According to research by Art Kramer, a psychologist at the University of Illinois in Urbana-Champaign, and others, from our mid-20s we lose up to one point per decade on a test called the mini mental state examination (*see* graph, below). This is a 30-point test of arithmetic, language and basic motor skills that is typically used to assess how fast people with dementia are declining. A three to four point drop is considered clinically significant. In other words, the decline people typically experience between 25 and 65 has real-world consequences.

That all sounds rather depressing, but there is an upside. The abilities that decline in adulthood rely on "fluid intelligence" – the underlying processing speed of your brain. But so-called "crystallised intelligence", which is roughly equivalent to wisdom, heads in the other direction. So even as your fluid intelligence sags, along with your face and your bottom, your crystallised intelligence keeps growing along with your

Cognitive ability, as measured by the battery of tests that make up the mini mental state examination, declines faster with every passing decade

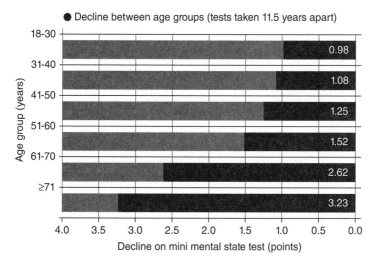

FIGURE 7.2 The decline in cognitive ability: Of a maximum score of 30, a decline of 4 points is considered clinically significant

waistline. The two appear to cancel each other out, at least until we reach our 60s and 70s.

Staying mentally and physically active, eating a decent diet and avoiding cigarettes, booze and mind-altering drugs seem to slow down the inevitable decline. And if it is too late to live the clean life, don't panic. You still have one more chance to turn it around.

Old age: Down but not out

By the time you retire, there's no doubt about it, your brain isn't what it used to be. You forget people's names and the teapot occasionally turns up in the fridge.

There is a good reason why our memories start to let us down. At this stage of life we are steadily losing brain cells in critical areas such as the hippocampus – the area where memories are processed. This is not too much of a problem at first; even in old age the brain is flexible enough to compensate. At some point, though, the losses start to make themselves felt.

Exercise can certainly help. Numerous studies have shown that gentle exercise three times a week can improve concentration and abstract reasoning in older people, perhaps by stimulating the growth of new brain cells. Exercise also helps steady our blood glucose. As we age, our glucose regulation worsens, which causes spikes in blood sugar. This can affect the dentate gyrus, an area within the hippocampus that helps form memories. Since physical activity helps regulate glucose, getting out and about could reduce these peaks and, potentially, improve your memory.

In fact, your brain is doing it all it can to ensure a contented retirement. During the escapades of your 20s and 30s and the trials of midlife, it has been quietly learning how to focus on the good things in life. By 65 we are much better at maximising the experience of positive emotion, says Florin Dolcos, a neurobiologist at the University of Alberta in Canada. In experiments, he found that people over the age of 60 tended to remember fewer emotionally negative photographs compared with positive or neutral ones than younger people.

MRI scans showed why. While the over-60s showed normal activation in the amygdala, a region of the brain that processes emotion, its interaction with other brain areas differed: it interacted less with the hippocampus than in younger people and more with the dorsolateral frontal cortex, a region involved in controlling emotions. Dolcos suggests that this may be a result of more experience of situations in which emotional responses need to be kept under control.

So it's not all doom and gloom. In fact you should probably stop worrying altogether. Studies show that people who are more laid back are less likely to develop dementia than stress bunnies. In one study, people who were socially inactive but calm had a 50 per cent lower risk of developing dementia compared with those who were isolated and prone to distress. This is likely to be caused by stress-induced high levels of cortisol, which may cause shrinkage in the **anterior cingulate cortex**, an area linked to Alzheimer's disease and depression in older people.

So while our brains may not wrinkle and sag like our skin, they need just as much care and attention. When you notice the signs of age, go for a walk, do a crossword and try to have a laugh – it might just counteract some of the sins of your youth.

Male brains, female brains

In recent years, numerous differences have been identified between men's and women's brains – some at the level of synapses, some in signalling chemicals and some in the size of particular brain regions. As a result, the idea that brains come in either pink or blue has gained a foothold in the public perception of neuroscience.

But there's a problem. Not all of these physical and physiological differences add up to differences in ability. Some researchers now believe that we have been overlooking something important. It is possible, they argue, that at least some of those disparities evolved not to create differences in behaviour or ability, but to prevent them. That the differences are there to compensate for the genetic or hormonal differences that are necessary to create two sexes with different sets of genitals and reproductive behaviours.

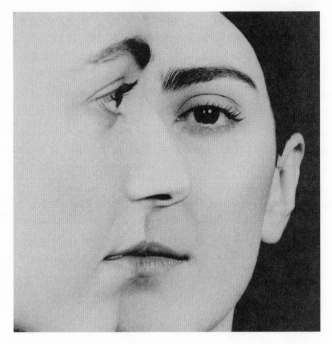

FIGURE 7.3 Looks like Venus and Mars aren't so far apart after all

If that sounds paradoxical, imagine comparing a chunky mountain bike with a lightweight road bike. To compensate for the mountain bike's greater resistance, you have to pedal harder to reach the same speed; one difference makes you introduce another to achieve the same output. In brain terms, while certain circuits may be shaded pink or blue that would not stop the output, or behaviour, being a uniform purple.

For most of history, men's and women's different roles in life were assumed to be mainly innate and unalterable. This was challenged in the West with the rise of feminism in the second half of the last century. Perhaps the different behaviours of boys and girls arose because of cultural norms: parents praising boys for romping and smashing toy cars, for instance, while expecting girls to be more reserved and play with their dolls.

Around the same time, though, new light was being shed on the biology of gender. In the womb, we all start out more or less female, until sometime between 6 and 12 weeks of pregnancy. Then, in male foetuses, a gene on the Y chromosome causes certain cells to make testosterone, which leads to the development of the penis and testicles. Female foetuses do not have this "testosterone bath" and so develop female reproductive organs.

But the sex hormones' influence is not limited to our gonads: they also play a key role in the brain's development, influencing the architecture of various neural circuits. As well as establishing these anatomical differences, the sex hormones presumably affect our behaviour as adults too, as their receptors have been found in many brain regions.

Understanding the ways in which male and female brains differ has become a hot topic in neuroscience, particularly in the past decade with the growth of brain scanning as a research tool. One of the most famous findings is that men seem to have a larger region of the brain thought to be involved in spatial reasoning, such as that used in a task like mentally rotating three-dimensional figures: the left-hand-side **inferior parietal lobule**, located just over the ear. Women, on the other hand, appear to have larger areas of the brain associated with language.

A common critique of this sort of work is that there is only a small average difference between the sexes, with more variability within each sex than between men and women as a whole. The results tell us about population averages, not individuals, in other words.

Nevertheless, there are differences in the numbers of women represented in different professions. For instance, women make up about 20 per cent of computer science students in the US, and the same fraction of engineering students. Is it down to innate brain differences or cultural conditioning that they miss out on these well-paying sectors so crucial to today's technology-oriented society?

Research into brain sex differences has also fuelled calls to educate boys and girls separately in same-sex classes or schools, particularly in the US. It is argued that teaching methods need to be tailored to those differently hued brains.

With such wide-ranging implications for society, it is important to be aware of possible flaws in the ideas about male and female brains. **Compensation theory** first caught people's attention in 2004 with Geert de Vries, who studies hormones and brain signalling systems in rodents at the University of Massachusetts in Amherst.

Insight from prairie voles

In the 1980s he stumbled across a big sex difference in the brains of prairie voles, small rodents found in the US Midwest. Unlike most mammals, prairie voles are monogamous and the males are devoted fathers. They spend just as much time as the females licking their pups and toting them around. Yet compared with the females, males have many more receptors in the brain for **vasopressin**, a brain signalling molecule that has been linked to parental care.

De Vries wondered if while female voles' maternal devotion was demonstrably triggered by the hormonal changes of pregnancy, the males' vasopressin circuits seemed to be compensating for the lack of pregnancy hormones. Soon afterwards he found several possible compensatory mechanisms in other animals, including rats, mice and zebra finches.

One convert to the idea is Margaret McCarthy, a sex differences researcher at the University of Maryland School of Medicine in Baltimore. "Many of the sex differences we see in the brain are there to help males and females develop their different reproductive strategies," she says. "But those differences also carry with them some constraints. Males have high testosterone; females have cycles of various hormones. And those hormones come with costs with regards to behaviours outside reproduction."

To date, the evidence for compensation in people seems thin on the ground. But could it be going unnoticed because of the assumption that a difference in the brain always means a difference in performance?

In a 2006 review of sex difference research, Larry Cahill at the University of California, Irvine, cited several brain-scanning studies that had turned up differences in men and women that were not accompanied by differences in their performance. While the mechanisms involved are unknown, Cahill thinks these could represent compensation in action, although they had not been noted as such by those who did the research.

Equal but different

Cahill himself may have found evidence of compensatory circuits at work, involving the amygdalae, a pair of almond-shaped structures deep within the brain thought to be involved in the processing and memory of emotional reactions. Cahill's group showed that even when the brain is at rest, amygdala activity is different in men and women. Cahill thinks the difference in amygdala activity could be a compensatory mechanism to make up for differences in testosterone levels.

Jill Goldstein's lab at Harvard Medical School in Boston did not go looking for a compensation effect; she believes De Vries's theory could explain her results.

Goldstein's team did fMRI scans on 12 women and 12 men as they viewed a variety of photos, some of which were designed to be shocking (think car accidents and dismembered bodies). The women did the test twice: once at the beginning of their menstrual cycle, when oestrogen levels would have been low and then again just before ovulation, when they would have been peaking.

When viewing the gruesome photos the women reported similar subjective feelings of stress as the men, irrespective of the stage in their menstrual cycle. But when their oestrogen was high, the women had less activity than men in several different brain regions involved in the stress response. Goldstein thinks this was to dampen down a more sensitive stress response that otherwise would have been triggered by the surging oestrogen. "They had the same subjective feelings of stress but their brains were acting slightly differently to get to that state," she says.

While the compensation theory has not yet gained much traction among neuroscientists, it is getting harder to ignore as the number of possible human examples accumulates. Even where compensatory brain differences have no net effect on behaviour or ability, they could still help explain why certain medical conditions are more common in one sex than the other. Women, for instance, are more vulnerable to mental illnesses like anxiety and depression, while men have a higher incidence of developmental disorders like autism.

Goldstein's work on stress is a case in point. "We need to understand how these circuits develop differently in the healthy male and female brain," she says. "Only then can we understand how these circuits are disrupted in psychiatric disorders."

Funders are also starting to take the issue seriously. In 2014 the US National Institutes of Health issued new policies requiring that sex differences be addressed in future biomedical research programmes funded by the agency.

No one is saying the compensation theory can explain away all the observable brain differences between men and women. Many of them do in fact correspond to differences in performance. But some do not.

That suggests we should be more careful about how we interpret brain data from now on, according to Lise Eliot, a neuroscientist at Rosalind Franklin University in Chicago, who coined a phrase with the title of her 2010 book on sex differences, *Pink Brain, Blue Brain*.

"The more we learn, the more we realise that sex differences don't translate very well into that Mars-Venus pop culture everyone seems to want to project," she says. "Neuroscientists, the media, parents – we all need to be careful about how this data is interpreted and what conclusions we draw from it."

What's it like to be a baby?

For those who want to get inside a baby's head, Alison Gopnik, a psychologist at the University of California, Berkeley, has a few suggestions: go to Paris, fall in love, smoke four packs of Gauloises cigarettes and down four double espressos. And then, she suggests, add psychedelic drugs. Because as far as we can tell, a baby's world is seriously strange.

Insights into what a baby's brain does have come from an increasing understanding of adult consciousness. Studies that pinpoint waves of electrical activity in the brain have shown that adults access information in two stages. The first involves unconscious processing of, say, an image. After some 300 milliseconds, the second stage kicks in, and a network of brain regions starts reverberating. This is the part when we become conscious of what we have perceived. Recent research has found that something similar happens in babies, but much more slowly. In babies from 12 to 15 months old, the second stage takes 750 milliseconds to come online; in five-month-olds, the lag was 900 milliseconds.

But there is more to an infant brain than just being a slower version of adult consciousness. Gopnik thinks that the kind of consciousness experience adults have is at one extreme of a spectrum. Babies are certainly not fully unaware but they are not conscious either. They may find themselves somewhere in the middle.

Philosopher Ned Block of New York University calls this **phenomenal consciousness** – what it's like to have a subjective experience such as seeing, hearing, tasting, smelling or touching something. When we observe a complex scene, we are conscious of a lot more than we can put into words.

Sensory bombardment

For a baby this scene may be all there is. Instead of paying attention to individual things, a baby is probably picking up patterns in the bombardment of stimuli. And because they are less able to control their attention, babies are drawn to things that are rich in information. For an adult, an infant's play area is a cacophony of colour and sound. A baby, however, is in its element.

This inability to control attention probably also means that babies struggle to shut things out. Adults might be able to tune out noises from outside while they are focusing on something else. A baby, though, is likely to find it hard to shut out the noise to begin with. As an infant, the world may be bright and brash, with no dimmer switch.

Which brings us to the bustle of Paris, being in love, and buzzing on coffee and cigarettes. The activation of certain parts of the brain for focused attention is managed by the neurotransmitter **acetylcholine**, which is mimicked by nicotine. At the same time, inhibitory neurotransmitters should work to stop other areas from joining the party. Unless you are drinking coffee, that is, because caffeine is thought to keep the effects of such killjoy neurotransmitters at bay, keeping your brain alert to anything and everything.

By smoking and drinking coffee you nudge your brain into a state where you're paying lots of attention – but in a wide-eyed, indiscriminate way. Being in love and travelling to new places seem to have a similar effect, says Gopnik. Under these influences, we get a more pliable, plastic brain. And that's a fair approximation of what's happening with babies, whose immature brains are more plastic overall. Being a baby is like paying attention with most of our brain.

It gets stranger. In infants, this expansive, along-for-the-ride experience of the world may go beyond perception. A baby's sense of self is mixed up with its awareness of other people. That means babies may feel their own emotions and the emotions of others, without being able to tell them apart.

The insight from mushrooms

The idea that infants might be experiencing an unbounded sense of self finds support from an unlikely source: magic mushrooms. And this also gives us another way to mimic infant consciousness. Robin Carhart-Harris of Imperial College London and his colleagues have been studying the effects of **psilocybin** – the active ingredient in psychedelic mushrooms – on states of consciousness. They looked at the network that connects regions in the prefrontal cortex, the cingulate cortex and the temporal lobes, among others.

Previous studies have shown that this "default mode network" is active when we are resting and when we are thinking about ourselves, and suppressed when we concentrate on a task. Carhart-Harris's team showed that psilocybin deactivates hubs in the brain like the posterior cingulate cortex and medial prefrontal cortex, as well as reducing long-range connectivity between brain regions. These hubs are like conductors of an orchestra, says Carhart-Harris. Bring on psilocybin – or stronger psychedelic drugs like LSD and the conductors leave the room.

Gopnik warns about the dangers of revisiting our weird and wonderful past. "LSD is dangerous, nicotine is very dangerous and nothing is more dangerous than falling in love," she says. "Tea with toddlers is really the safest way to expand your consciousness."

8
Sleep

Most people think they don't get enough of it. Some manage just fine on half the usual amount. Dolphins do it with half a brain at a time. Rats die after three weeks without it, yet male Emperor penguins manage an entire three months of brooding without getting any. It's as good as ubiquitous among animals – even insects do it. We spend a third of our lives at it. Yet sleep is still one of life's great enigmas.

What is sleep?

Strictly speaking, the term "sleep" only applies to animals with complex nervous systems. Yet its origins might date back to the dawn of life about 4 billion years ago, when the earliest microorganisms changed their behaviour in response to night and day. We know that today's microorganisms, even without anything resembling a nervous system, have daily cycles of activity and inactivity driven by internal **circadian clocks**.

Invertebrates, such as scorpions, insects and some crustaceans also have sleep-like states in which they go through cycles of rest and activity. When resting they take up a stereotypical body position, stop responding to the outside world, and if you rouse them they later compensate for what they have lost.

Because of this, some researchers consider sleep to be part of a continuum of inactive states found throughout the animal kingdom.

Why we and other animals spend so much time sleeping is still largely a mystery. Once we understand more about what benefits an animal gets from its sleep-like state, we may be able to provide a meaningful answer to the question of what sleep is and what it is really for.

Cycles of activity

Human sleep is accompanied by complex changes that occur in the brain. This can be observed with an electroencephalogram (EEG), which measures the brain's electrical activity over time.

31% of drivers in the US report having fallen asleep at the wheel at some point in their lives

7 to 9 *hours of nightly shuteye is best for adults*

After lying awake for ten minutes or so we enter non-rapid eye movement sleep or NREM sleep. NREM sleep is divided into three stages, NREM1, NREM2 and NREM3, based on subtle differences in EEG patterns. Each stage is considered progressively "deeper".

After cycling through the NREM stages we enter rapid-eye-movement or REM sleep. The EEG pattern during REM sleep is similar to wakefulness or drowsiness. It is during this stage that many of our dreams occur.

Each cycle lasts for about 1.5 hours and a night's sleep usually consists of five or six cycles.

In addition to changes in brain activity, sleep is also characterised by a reduction in heart rate of about 10 beats per minute, a fall in core body temperature of 1°C to 1.5°C, as well as a reduction in movement and sensation.

Why we sleep

Why we sleep is one of life's great enigmas. It's clearly crucial to our survival: when lab rats are deprived of sleep, they die within a month, and when people go for a few days without sleeping, they start to hallucinate and may have epileptic seizures.

An explanation of why has not been easy to come by, however. If our bodies just needed physical rest, for example, we could do this without falling unconscious. "Why would we spend several hours a day at the mercy of predators without

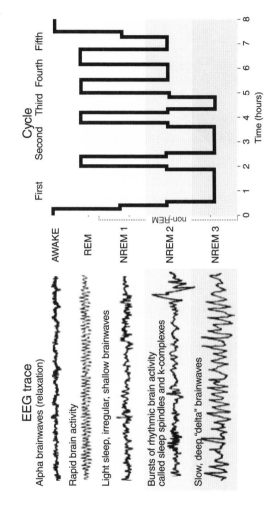

FIGURE 8.1 Human slumber cycles

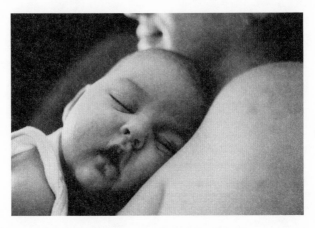

FIGURE 8.2 Why do babies spend the majority of their time asleep?

doing some useful function?" asks Giulio Tononi at the University of Wisconsin–Madison.

Most theories involve some kind of brain housekeeping or repair, but until recently no compelling candidates had emerged for what changes inside the brain during sleep.

Recently, a "drainage system" was discovered, similar to the lymph vessels that drain waste products from tissues in the rest of the body.

This ramps up when we sleep, suggesting that waste products are being cleared away while we slumber.

Alternative theories suggest the purpose of sleep is to conserve energy or keep animals out of harm's way while food is unavailable. And of course it is possible that sleep could be performing several functions at once, or may even have different purposes in different animals.

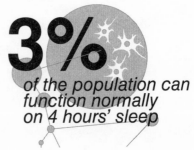

3% of the population can function normally on 4 hours' sleep

Tononi, though, suggests that a major function of sleep is to stop our brains becoming overloaded with new memories. The idea is that sleep evolved so that connections between neurons in the brain can be whittled down overnight, making room for fresh memories to form the next day.

11 days the longest anyone has stayed awake

In recent experiments, Tononi's team measured the size of these connections, or synapses, in the brains of 12 mice. The synapses in samples taken at the end of a period of sleep were 18 per cent smaller than those in samples taken before sleep, showing that the connections between neurons weaken during slumber. Because memories are stored by new synapses, and sleep seems to consolidate new memories, the idea that sleep weakens, rather than strengthens, brain connections might seem counterintuitive. But there is previous evidence to support the housekeeping theory. For instance, EEG recordings show that our brains are less electrically responsive at the start of the day – after we have slept well – than at the end, suggesting that the connections may be weaker.

Tononi's research also hints at how we may build lasting memories over time, despite this thinning. The team discovered that some synapses seem to be protected – the biggest fifth stayed the same size. It's as if the brain is preserving its most important memories, he says. "You keep what matters."

If the housekeeping theory is right, it would explain why, when we miss a night's sleep, we find it harder the next day to concentrate and learn new information – we may have less capacity to encode new experiences. Tononi's finding also suggests that, as well as sleeping well after learning something new, we should also try to sleep well the night before. "Sleep is the price we pay for learning," he says.

Is it possible to be half-asleep?

Sleep feels like an on-or-off condition, but brains can be awake and asleep at the same time. This phenomenon is well known in dolphins and seals – animals that can sleep "uni-hemispherically": one half of their brain is asleep while the other half shows electrical activity characteristic of wakefulness.

Sleep researchers wondered if something similar happens in humans. Is sleep a "global" state that happens to the whole brain, equally, or can it, to some extent, be regulated locally? There is mounting evidence for the latter. For example, the most active brain regions during wakefulness subsequently undergo deeper sleep for longer.

This localised view of sleep could lead to a better understanding of cases when wakefulness intrudes into sleep, such as in sleep-talking, sleepwalking and a rare form of insomnia in which people report being awake all night even though recording brainwaves from a single brain area suggests they have been asleep.

It may also explain how sleep can intrude into wakefulness, such as during lapses of attention when we are sleep-deprived. These "micro sleeps" can be particularly dangerous when driving and various ways to detect them have been developed, for instance by monitoring how a car moves relative to white lines on roads or analysing the movements of the eyes for signs of sleepiness.

Why do we dream?

"The interpretation of dreams is the royal road to a knowledge of the unconscious activities of the mind." So wrote Sigmund Freud in his 1900 classic, *The Interpretation of Dreams*. He saw this idea as a "once in a lifetime" insight, and for much of the 20th century the world agreed. Around the globe, and upon countless psychoanalysts' couches, people recounted their dreams in the belief that they contained coded messages about repressed desires. Dreams were no longer supernatural communications or divine interventions – they were windows into the hidden self.

Today we interpret dreams quite differently, and use far more advanced techniques than simply writing down people's recollections. In sleep laboratories, dream researchers hook up volunteers to EEGs and fMRI scanners and awaken them mid-dream to record what they were dreaming. Still tainted by association with psychoanalysis, it is not a field for the faint-hearted. "To say you're going to study dreams is almost academic suicide," says Matt Walker at the University of California, Berkeley.

Nevertheless, what researchers are finding will make you see your dreams in a whole new light. Modern neuroscience has pushed Freud's ideas to the sidelines and has taught us something far more profound about dreaming. We now know that this peculiar form of consciousness is crucial to making us who we are. Dreams also help us to consolidate our memories, to make sense of our myriad new experiences and keep our emotions in check.

100,000
the number of annual car crashes in the US related to fatigue

Changing patterns of electrical activity tell us that the sleeping brain follows 90-minute cycles, each consisting of five stages (*see* Figure 8.1 above). There is no characteristic pattern of brain activity corre-

33% *of dreams contain bizarre elements impossible in everyday life*

sponding to dreaming, but as far as we know all healthy people do it. And while dreaming is commonly associated with REM sleep, during which it occurs almost all of the time, researchers have known since the late 1960s that it can also occur in non-REM sleep – though these dreams are different. Non-REM dreams tend to be sparse and more thought-like, often without the complexity, length and vivid hallucinatory quality of REM dreams.

Despite their differences, both types of dreams seem to hold a mirror to our waking lives. Dreams often reflect recent learning experiences and this is particularly true at the start of a night's sleep, when non-REM dreaming is very common. Someone who has just been playing a skiing arcade game may dream of skiing, for example. The link between waking experience and non-REM sleep has also been observed in brain scanning studies. Pierre Maquet at the University of Liège, Belgium, looked at the later stages of non-REM sleep and found that the brains of volunteers replayed the same patterns of neural activity that had earlier been elicited by waking experiences. Many REM-sleep dreams also reflect elements of experiences from the preceding day, but the connection is often more tenuous – so someone who has been playing a skiing game might dream of rushing through a forest or falling down a hill.

Sleep on it

35% of US workers sleep fewer than 7 hours a night

But we do not simply replay events while we dream, we also process them, consolidating memories and integrating information for future use. Robert Stickgold of Harvard Medical School in Boston recently found that people who had non-REM dreams about a problem he had asked them to tackle subsequently performed better on it. Likewise, REM sleep has been linked with improved abilities on video games and visual perception tasks, and in extracting meaning from a mass of information.

"It's clear that the brain does an immense amount of memory processing while we sleep – and it certainly isn't mere coincidence that while our brain is sorting out these memories and how they fit together, we're dreaming," says Stickgold. He suspects that the two types of dream states have different functions for memory, although what these functions are is a matter of debate. Non-REM dreaming might be more important for stabilising and strengthening memories, Stickgold suggests, while REM dreaming reorganises the way a memory is stored in the brain, allowing you to compare and integrate a new experience with older ones.

However, Jan Born and Susanne Diekelmann, now both at the University of Tübingen in Germany, have looked at the same evidence and come to the opposite conclusion – that REM sleep supports the strengthening of a new memory, while non-REM sleep is for higher-level consolidation of memories. "I think this means that we're still lost when it comes to understanding the role of different sleep stages in memory," says Stickgold.

Also unclear is how central the role of dreams in memory formation is. The dreaming phase is certainly not the only time our brains consolidate memories. For example, when we daydream certain areas of the brain, called the **default network**, become active. We now know this network is involved in memory processing and many of the same brain regions are active during REM sleep. What's more, daydreaming, like REM dreaming, can improve our ability to extract meaning from information and have creative insights.

Does this mean we don't actually need dream sleep to process memories? Not necessarily, says Walker, who points out that the way new memories are replayed in the brain is different in daydreaming and dreaming. Rat studies show that the reruns happen in reverse when the animals are awake and forwards when they are sleeping. No one is quite sure what this difference means for memory processing, but Walker believes it shows that daydreaming is not simply a diluted version of sleep dreaming. Maquet agrees. "Different brain states may all have somewhat different functions for memory. Memory consolidation is probably organised in a cascade of cellular events that have to occur serially," he says. Some while you are awake, and then some while you are asleep.

Emotional reset

Even if dreaming is crucial for memory, Walker for one does not see this as its main function. "I think the evidence is mounting in favour of dream sleep acting as an emotional homeostasis: basically, rebalancing the emotional compass at the biological level," he says. Every parent knows how a short nap can transform a cantankerous two-year-old and Walker has shown something similar in adults. He found that a nap that includes REM dreaming mitigates a normal tendency in adults to become more sensitive to angry or fearful faces over the course of a day, and makes people more receptive to happy faces.

Walker has also found that sleep, and REM sleep in particular, strengthens negative emotional memories. This might sound like a bad thing – but if you don't remember bad experiences you cannot learn from them. In addition, both he and Stickgold think that reliving the upsetting experience in the absence of the hormonal rush that accompanied the actual event helps to strip the emotion from the memory, making it feel less raw as time goes on. So although dreams can be highly emotional, Walker believes they gradually erode the emotional edges of memories. In this way, REM dreams act as a kind of balm for the brain, he says. In people with post-traumatic stress disorder, this emotion-stripping process seems to fail for some reason, so that traumatic memories are recalled in all their emotional detail – with crippling psychological results.

As with memory processing, REM and non-REM dreaming may play different psychological roles. Patrick McNamara of Boston University has found that people woken at different sleep stages give different reports of their dreams. REM dreams contain more emotion, more aggression and more unknown characters, he says, while non-REM dreams are more likely to involve friendly encounters. This has led him to speculate that non-REM dreams help us practise friendly encounters while REM dreams help us to rehearse threats.

What do dreams mean?

All of this suggests that we couldn't function properly without dreaming, but it doesn't answer the perennially intriguing question: what do dreams actually mean?

For some sleep researchers, the answer is simple – and disappointing. Born argues that dreams themselves have no meaning, they are just an epiphenomenon, or side effect, of brain activity going on during sleep, and it is this underlying neuronal activity, rather than the actual dreams, that is important. Walker finds it hard to disagree. "I don't want to believe it. But I don't see large amounts of evidence to support the idea [that dreams themselves are significant]," he says.

Those researchers who refuse to accept the notion that the content of dreams is unimportant point to work by Rosalind Cartwright of Rush University, Chicago. In a long series of studies starting in the 1960s, she followed people who had gone through divorces, separations and bereavements. Those who dreamed most about these events later coped better, suggesting that their dreams had helped. "Cartwright's work provides some of the most solid evidence that dreaming serves a function," says Erin Wamsley at Furman University in Greenville, South Carolina.

In fact, Wamsley's own research hints that the form and function of a dream are connected. She worked with Stickgold on the study that found that non-REM dreams boost people's performance on a problem. Their volunteers were given an hour's training on a complex maze, then either allowed a 90-minute nap or kept awake. The dreamers subsequently showed bigger improvements, but the biggest gains of all were in people who dreamed about the maze. It didn't seem to matter that the content of these dreams was obtuse. One volunteer, for example, reported dreaming about the maze with people at checkpoints – although there were no people or checkpoints in the real task – and then about bat caves that he had visited a few years earlier. Stickgold didn't expect this to improve the volunteer's ability to navigate the maze, "and yet this person got phenomenally better".

He points out that the dream content is consistent with the idea that during dreaming memories are filed with other past experiences for future reference. "Dreams have to be connected in a meaningful, functional way to improvements in memory – not just be an epiphenomenon," he says. "I say this with fervent emotion, which is what I use when I don't have hard data."

Such evidence may one day be forthcoming, though. In the past, there has been no objective way to record what someone is dreaming, but that is changing. In 2008, Yukiyasu Kamitani at the ATR Brain Information Communication Research Laboratory in Kyoto, Japan, and colleagues used fMRI scans to decode and then recreate scenes that volunteers were picturing in their mind while awake. To see if they could do the same thing with people's dreams, in a later study the team repeatedly woke volunteers as they slept in a scanner and asked them to describe their dreams. Using that information, they were able to categorise what certain patterns on fMRI scans meant and tell with 60 per cent accuracy what kinds of things people were dreaming about – for instance whether they were dreaming about a man or woman, or certain types of objects, such as a car.

Some may think all this peering and prodding at our dream world is taking away its magic, but the researchers don't see it that way. While you are dreaming, your brain literally reshapes itself by rewiring and strengthening connections between neurons. So although dreams do not reveal the secret you, they do play a key role in making you who you are. "The mystery and the wonder of dreams is untouched by the science," says Stickgold. "It just helps us appreciate better how amazing they really are."

Sleep: a user's guide

It should be obvious by this point that sleep is crucial to our brain's health and emotional stability. But how much do we need to get these magical effects? Here, too, scientists are hard at work looking for answers.

How much do I need?

We all know eight hours is the magic number for a decent night's sleep. The catch is that nobody seems to know where this number came from. In questionnaires, people tend to say they sleep for between seven and nine hours a night, which might explain why eight hours has become a rule of thumb. But people also tend to overestimate how long they have been out for the count.

According to Jerome Siegel, who studies sleep at the University of California, Los Angeles, the eight-hour rule has no basis in our evolutionary past – his study of tribal cultures with no access to electricity found that they get just six or seven hours with no obvious health problems.

So perhaps eight hours is the wrong target and we can get by just fine with seven. This seems to be a minimum requirement. A recent analysis in the US concluded that regularly getting less sleep than that increases the risk of obesity, heart disease, depression, and early death, and recommended that all adults aim for at least seven hours.

By this benchmark, recent reports seem to suggest we are walking around in a state of sleep deprivation. The US Centers for Disease Control and Prevention estimates that 35 per cent of US adults are getting less than seven hours a night, and a survey in the UK found that the average was 6.8 hours. The media widely state that we are getting less sleep than we used to. The implication is that it's taking a severe toll on our health.

Not everyone is convinced. "Sleep has not changed in the past 100 or so years," says sleep scientist Jim Horne, who takes the idea to task in his book *Sleeplessness – Assessing sleep need in society today*. That's a notion backed up by a recent review of scientific literature on sleep between 1960 and 2013, which found no significant link between sleep duration and the year a study was conducted.

What studies have shown is that the amount of sleep we need is influenced by our genes and varies among individuals. Exactly which genes are involved is not well understood, but a recent study of more than 50,000 people found one gene variant that added 3.1 minutes of sleep for every copy you have. The amount of sleep you need also changes as you age.

Taking this into account, the US National Sleep Foundation updated its guidelines recently, and came up with a recommended range of seven to nine hours for adults, but with added leeway of an hour either side to account for natural variation (*see* Figure 8.3 below).

What of those grating individuals who claim to get by just fine on a few hours each night? They probably are sleep-deprived, but have got used to the effects and now fail to notice them as strongly. Or else they may simply be napping later on in the day. Only a tiny minority of us, probably less than three per cent, can get by on four to six hours of sleep with no problems at all. Ying-Hui Fu at the University of California, San Francisco, and her colleagues found a particular gene in

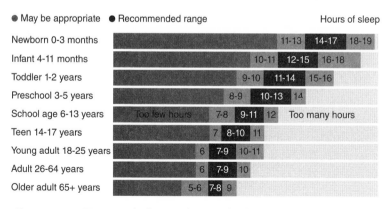

FIGURE 8.3 How much shuteye do I need? Sleep need varies as we age, and for each individual.

a family of these natural short-sleepers. When the team engineered mice to express this short-sleep gene, they recovered from sleep deprivation quicker and seemed to whizz through the non-REM stages of sleep faster than non-engineered mice.

The team thinks this gene variant interacts with proteins that are at the core of the circadian clock, opening up the tantalising possibility that we could one day genetically engineer our way to a shorter night's sleep, without the downsides. In the meantime, it is comforting to know that, for most of us, getting stuff done on very little sleep is, so far, physically impossible.

Can you sleep too much?

Getting more time under the covers asleep can be good for your health, but beware. It seems that you can indeed have too much of a good thing.

Shawn Youngstedt of Arizona State University in Tempe has studied epidemiological data, and these suggest that regularly getting eight hours or more could send you to an early grave. "Typically, the association is at least as strong, often stronger, than the association of short sleep with mortality," he says. "There seems to be a sweet spot for all health-related behaviours. For sleep that seems to be about 7 hours."

Just why this is remains a mystery. It could come down to the simple fact that when we are asleep we are moving very little, and there's plenty of evidence to show that inactivity is bad for you.

And although this might not matter if you are active during the day, it could be that people who spend more time asleep do less exercise, possibly because they simply have less time, Youngstedt suggests. Long sleep is also

---

associated with inflammation, an immune response linked to everything from depression to heart disease. And you might not need as much sleep as you think, says Young-stedt. "Many sleep for a long time out of habit or boredom, and we have found that they can tolerate mild sleep restriction," he says. So try cutting down. You might even feel less tired.

9
Technology to improve your brain

Connecting your brain directly to electronics, bypassing the muscles and senses that normally connect us to the world, was once the realm of science fiction. It is now rapidly becoming a clinical reality.

Neural implants

Our ability to monitor and manipulate electrical signals in the brain is growing by the day, opening new possibilities to restore functions lost through injury or disease. What started with cochlear implants has advanced to deep brain stimulation and artificial retinas and even to brain implants that enable control of devices outside of the human body.

Broadly speaking, neural implants have the potential to do one of two things. Input devices send electrical signals directly into the nervous system. These can relay sensory information from the outside world, such as sound or light, as in cochlear implants and artificial retinas. Input devices can also be used to control the

---

Cochlear implants

The first cochlear implant, designed by American surgeons John Doyle and Williams House, was tested in a human volunteer in 1964. At first, arrays of four electrodes positioned in the inner ear gave just enough detail to allow patients to hear and repeat simple phrases.

Today, commercially available implants with more than 20 electrodes restore hearing to tens of thousands of people every year. In particular, children receiving cochlear implants at an age when their brains are plastic enough to make sense of the new inputs can achieve remarkable speech comprehension.

Now researchers are working on totally implantable devices with no external wires or batteries, as well as exploring ways to improve performance with more complex sounds such as music.

---

misfiring of the brain that causes epileptic seizures or the tremors of Parkinson's disease. By contrast, output devices send electrical signals from the brain to the outside world, for example to move external devices such as prosthetic limbs. Input devices were the first to be put to use in humans, such as cochlear implants.

Technologies are also being developed to stimulate the retina and restore vision. Retinal implants such as "Second Sight", which was first trialled in patients in 2002, electrically stimulate surviving cells in the eyes of people with blindness caused by damage or loss of light-sensitive parts of the retina. Images captured by cameras are translated into a crude version of what a healthy eye would see. Because of the richness of the visual information we receive from the eyes, it's quite a challenge to make something as good as normal sight, but scientists are working on improved implants with thousands of individual pixels.

Input devices that stimulate the brain directly have a similarly long history. A pioneer was José Delgado, a Spanish researcher who

FIGURE 9.1 Thought-controlled wheelchair

later did much of his work in animal models at Yale University in the 1950s and 60s. He developed a brain implant that he called the stimoceiver that he attached to specific regions of animals' brains to try to control their behaviour. In one famous experiment (video of which is easy to find online) he sent electrical pulses via remote control to the brain of a charging bull and stopped it in its tracks. Delgado even went on to try his stimoceiver in the brains of human volunteers.

Optogenetics

In future it may be possible to control brain activity using light rather than electricity. Optogenetics is a new technology whereby neurons can be genetically engineered to respond to light. A first application could be to restore light-sensitivity to the retina, but in theory the technique can also be used in the brain if we can find a safe way to deliver the gene therapy and the light inside the skull.

Experiments in animals have shown that optical stimulation can be used to activate or silence specific neurons and influence behaviour, for example making mice run when light is shone on motor areas of the brain. When the light stops, so do they. In the future, this ability to optically control neurons could be extremely useful for conditions such as epilepsy.

Optogenetic silencing of neurons has already been shown to suppress seizures in mice. Researchers are now working on a device that will monitor brain activity to detect when a seizure starts. Combined with gene therapy this could one day allow us to turn off seizures as quickly as they start.

Fast forward a few decades and the applications of this kind of technology are becoming clear. More than 100,000 people worldwide have received deep brain stimulation implants designed to tame the tremors associated with Parkinson's disease. Electrodes inserted into the **basal ganglia**, the part of the brain that contains dopamine-producing neurons that are lost in Parkinson's disease, are linked to a battery implanted in the chest and controlled by a handheld remote. When turned on, the tremors stop completely and users are free to go about their daily life.

More recently, output devices have been developed – those that record signals from the brain. These have enabled scientists to read neural activity and translate it into signals that can be used to control prosthetics. By placing fine electrodes in the motor cortex it is possible to identify 'spikes' that correspond to activity of single neurons. Early experiments in monkeys showed that by listening to the patterns of spikes in multiple neurons it was possible to "decode" the direction of arm movements. The next step was to use that information to move a robotic arm which monkeys quickly learned to control using only the brain signals. In the past decade it has become possible to do something similar in people.

In 2004 Matthew Nagle, who was paralysed from the neck down, became the first person to have a brain implant that enabled him to control a cursor on a computer screen, and to open and close a hand prosthesis. A number of other people have had similar implants since then, but the research is very much still in the lab – for a start it requires the person to be wired to a computer that translates their brain signals to the prosthetic. But with advances in low-power microelectronics, the next generation of neural interfaces are being designed to operate wirelessly.

Milestones in fighting paralysis

Combining brainpower with electronics is rapidly rolling back the limitations of paralysis. Electrodes in a locked-in woman's brain have allowed her to communicate by thought alone (*see* "First home brain implant lets 'locked-in' woman communicate"), but electrodes have also helped a partially paralysed man to recover some use of his hand. It was announced in April 2016 that linking the motor cortex area in his brain to an electrode sleeve on his arm allowed him to pour liquid from bottles and play *Guitar Hero*.

Electrodes can also restore a sense of touch. A study in October 2016 reported that a quadriplegic man was able to feel as if he was touching objects via a robotic arm, thanks to electrodes in his somatosensory cortex. But brain implants are not the only show in town. Caps of electrodes, worn on a person's head, have enabled some paralysed people to walk. The caps transmit brain signals to an exoskeleton, which is worn a bit like a pair of trousers and moves when signalled. An exoskeleton isn't always needed. In 2015, a paralysed man learned to walk without one, thanks to an EEG cap that sent signals to muscle-stimulating electrodes implanted in his legs.

Two-way communication

Recently, scientists have started working on **bidirectional interfaces** that combine input and output capabilities. Such implants could act as artificial connections to relay information between two parts of the nervous system that have been disconnected by injury: for example, testing a bidirectional interface connecting the brain and the spinal cord in monkeys using a drug to disrupt the motor cortex, temporarily mimicking the effect of a stroke and paralysing the monkey's hand. By connecting neurons upstream of the inactivated area to electrical stimulation in the spinal cord, it is possible to restore the monkey's ability to grasp with the hand. It has even been suggested that bidirectional implants could be used to restore cognitive functions such as memories, by replacing the input and output connections of the hippocampus, the brain area crucially involved in forming memories.

Another intriguing application of bidirectional neural interfaces is to change the strength of connections between brain areas, a phenomenon known as **neuroplasticity**. This is what healthy brains do naturally: every day the things we learn are reflected in the changing connections between signals that come in from the world and the signals we send back out. It's this interplay that drives the physical changes in the brain that makes learning happen. With bidirectional neural interfaces we can alter the relationships between inputs and outputs and thereby change the brain.

This is based on an idea that came from neuroscientist Donald Hebb back in the 1940s and is often described as "neurons that fire together wire together". It means that when two brain networks have been active together, the neural connections between them strengthen. Since a bidirectional interface can artificially link inputs and outputs, it should lead to stronger brain connections between areas.

In experiments in monkeys it has been shown that this does indeed work. Sites in the cortex that represented different movements of the arm (for example, flexing or extending the wrist) were linked. For several days two of the sites were connected using a bidirectional interface. Afterwards, a new movement representation in the brain was found that suggested a new pathway had been created that linked the sites.

In fact it is becoming increasingly clear that neural implants can exploit neuroplasticity in various ways. When the brain learns to control a robotic arm using a neural implant, it is effectively acquiring a new skill similar to when we learn to use a tool. Even if the neurons controlling the robotic arm would not normally move the real arm in exactly that way, the brain is smart enough to work out how to do it based on observing the effects. Indeed, if monkeys are shown a computer cursor that is moved by a single neuron, it takes only a few minutes for monkeys to learn to increase and decrease the activity of that neuron to reach different targets. This remarkable ability of the brain to learn through "neurofeedback" is a valuable capacity to exploit as we start to wire devices into the brain.

Once we start thinking of neural implants as tools, we are in theory not limited to simply replacing a lost function. One day we could enhance functions, or add new functions. Perhaps the brain could learn to connect directly to computers and access information from the worldwide web without needing a keyboard and computer screen. Or perhaps we could even connect to other people's brains and communicate our thoughts directly.

So far, this is all very much still science fiction, but in our favour we're plugging into the smartest learning machine in the known universe. Our ability to design and exploit a wide range of tools with our hands has driven the evolution of the human species. Perhaps we are now starting to witness the next stage in that evolution as we connect our brains directly to technology.

First home brain implant lets "locked-in" woman communicate

In 2016 a paralysed woman became the first person able to use a brain-computer interface at home in their daily life. "It's special to be the first," Hanneke De Bruijne told

New Scientist. De Bruijne has amyotrophic lateral sclerosis (ALS) – a disease that ravages nerve cells. "All muscles are paralysed. I can only move my eyes," says De Bruijne.

"She is almost completely locked in," says Nick Ramsey at the Brain Center Rudolf Magnus of the UMC Utrecht in the Netherlands. When Ramsey met De Bruijne, she relied on an eye-tracker to communicate. Using this, she can choose letters and spell out words on a screen. This works for now, but one in three people with ALS loses the ability to move their eyes. Teams around the world have been working on systems that are controlled directly by the brain instead.

These devices read brain activity and translate it into signals that can control a computer or a robotic limb, for example. But, so far, none of them fits easily into people's lives at home. Ramsey's team's interface uses electrodes placed on the surface of the brain, just underneath the skull. When brain activity is recorded by an electrode, a signal travels through a wire to a small device, which can be implanted under the skin of the chest, like a pacemaker. This implant wirelessly sends a signal to a tablet computer, which can transform it into a simple "click". Other software on the tablet permits the click to be used for various things, such as playing a game or using a speller to select words and communicate.

De Bruijne volunteered to have the system implanted last year. "I want to contribute to possible improvements for people like me," she says. The team inserted an electrode over a brain region that controls movement of the right hand. After multiple training sessions, which involved playing games such as whack-a-mole and Pong,

De Bruijne learned to trigger a click by imagining moving her hand. After six months, her accuracy was 95 per cent.

Next, Ramsey's team wants to develop software that can translate clicks into other functions. "My dream is to be able to drive my wheelchair," says De Bruijne.

Memory implants: can chips fix broken brains?

Decoding brain signals to hack into our senses and movements is one thing. But memories? Surely that's the stuff of science fiction. Sam Deadwyler's work certainly sounds like something from *The Matrix*. In the same way that Neo downloads a kung fu master's skills, Deadwyler had wired up the brain of a rat with electronics that transplanted memories derived from 30 rats into its brain, allowing it to draw on training that it had never personally experienced. The study had the potential to be a landmark finding – but "everyone thought it was science fiction," he says. "I thought, 'no one's going to believe this unless I do a hundred control experiments'."

So he did just that. And in 2013, ten years after the original experiment – the paper was finally published. Instant kung fu is still the stuff of Hollywood blockbusters, but this research could nevertheless have a huge impact on many people living with brain damage. The same kind of neural implants that allowed memories to be "donated" from many rats into another individual could restore lost brain function after an accident, a stroke or Alzheimer's disease.

For a lot of people with memory loss, damaged parts of the brain are failing to pass information from one area to another. If you could create electronics that interpret the signals from one area, circumvent the damaged parts, and write them into the second area, you could help people regain the ability to form new memories, or even gain access to precious old ones. Such a chip would act as a kind of brain bypass.

Getting there won't be easy: such an implant requires neuroscience that is only now beginning to be understood. More than that, however, these new technologies raise ethical questions that were once the preserve of science fiction. Our memories define us, so conserving them from damage could save our identity – but when your memory is a computer algorithm, are you still the same person? It's almost time to find out: the first human studies are underway.

Our ability to send messages into and out of the brain has accelerated rapidly over the past two decades, (*see* preceding pages and timeline).

Memory, however, is complicated by the fact that it requires coordination between multiple brain regions. Translating between different areas of the brain requires a device that can record activity from one set of neurons and then electrically stimulate another set of neurons to replay it whenever it is needed. Needless to say, it's an endeavour rife with challenges. "To make a cognitive device, we first have to know what a memory looks like," says Robert Hampson, who works on cognitive implants with Deadwyler at Wake Forest Baptist Medical Center in North Carolina. The search for a memory trace in the brain has been complicated by the fact that there are many different kinds of memories: there's the short-term "working memory" that helps you to remember a phone number before you dial, sense memories that might include the echo of what someone's just said and long-term memories of facts, skills and experiences. It is this long-term recall, and how it emerges from working memory, that Deadwyler and Hampson are interested in.

Although each memory trace is different, all long-term memories begin life in a region called the hippocampus, the brain's "printing press". Place an implant here and it might be possible to record memories as they form. The next step is to figure out the neural code that represents a particular memory. The key is thought to lie in the exact firing pattern of interconnected neurons; one synchrony of neurons might translate to your idea of the Eiffel Tower, for instance, while another, perhaps overlapping, network might represent Paris more generally.

Cracking the neural code

Quietly, over the past couple of decades, neuroscientists have begun to find ways to crack that code. Early steps were made in the 1990s by Theodore Berger at the University of Southern California, who turned to a technique called multi-input/multi-output (or MIMO). MIMO is more typically used to tease signals from noise in wireless communications, but Berger realised he could apply the same principle to pick out meaningful signals from the noise of millions of neurons firing. The quest didn't endear him to sceptics. "People called him crazy for a long, long time," says Eric Leuthardt, a neurosurgeon at the University of Washington in St Louis, Missouri.

It's not that memory signatures have been invisible to other scientists. There has been some evidence to suggest that people have extremely specialised neurons that fire in response to a single concept, such as their grandmother or Jennifer Aniston.

However, these so-called grandmother cells encode a narrow range of ideas, whereas Berger was chasing the ability to encode any memory. Slowly but surely he has shown that the MIMO algorithm can do this — by using it to isolate the specific signal behind the memory of an action, and then replaying that exact sequence.

In one ground-breaking experiment, Berger — now working with Deadwyler and Hampson — implanted a chip containing electrodes into the hippocampus of rats. Then they used the MIMO algorithm to isolate and record the relevant neural code as trained animals pressed one of two levers to receive a treat. After drugging the rats to impair their ability to remember which lever gave the treat, the team then used the same electrodes to deliver the same firing pattern back to the neurons. Despite its amnesia, the rodent knew which lever to push. In other words, the algorithm had helped to restore the rat's lost memory.

It was a triumph, demonstrating that electronics can crack the neural code and potentially replace damaged areas of the brain – acting as an artificial hippocampus to treat amnesia, for instance.

One key question raised by Deadwyler's research was whether we each have a different neural code, or whether there is a more generalised language shared by everyone. This is where Deadwyler and Hampson's attempts to transplant memories between rats come in.

Their experiments typically involved two sets of rats trained to run between two arenas, pressing a series of levers in a certain order. Importantly, one set was trained to delay their actions – they had to pause for up to 30 seconds before they were able to press one of the levers – while the second set was not. Presented with the unexpected delay, the second set of rats lost the plot – they could not remember which lever they had been taught to press. But when Deadwyler and Hampson used MIMO to record the brain activity for this task in the first group and replayed it in the second using electrodes, those rats began to act as if they had taken the alternative form of training, choosing the correct lever even after a long pause, even though that had not been part of their previous experiences. "Our model lets us establish a memory that has not been used before," Hampson says.

How to read a thought

We've come a long way in our ability to decode the meaning of the brain's signals. But before you can tease out what these signals might mean, you first need a high-fidelity signal. There are many ways to eavesdrop on the brain's electrical communications, and they're all about trade-offs.

Think of it as a night at a concert. Non-invasive electrodes on the scalp can listen to the entire orchestra. You can zero in on the string section by getting a little more invasive with electrocorticography, which involves placing a sheet of electrodes on the surface of the brain. But if you want to listen to an individual violin, you have no choice but to make direct contact with individual neurons. Such high precision requires inserting deep-penetrating electrodes – which come with several caveats. Their insertion can rip and slice the surrounding neurons, causing oedema and scarring, and the brain's immune response forms scars that ward off the invading object. Soon, the electrodes can no longer read any meaningful information. As a result, few implants have lasted more than a year or two. To get around this problem, neuroscientists are developing new types of electrodes that are more compatible with the body, or that can hide from the brain's immune system.

Was the set-up really this good? Or could their success be explained by some other cause? Deadwyler and Hampson embarked on an enormous number of control experiments to rule out every other explanation, including the possibility that it was just an unintentional artefact of electronically tickling the brain, or some general improvement caused by electrical stimulation. Finally, in December 2013 the paper was published:

it really is possible to plant a general signature of a memory in the brain.

If that could be replicated in humans, a chip could come with ready-made code that could give people a head start on relearning general skills, such as brushing your teeth or driving a car, say – actions that are often lost after brain damage. "Before we can get someone with brain damage back to work, we want to return their capability to form those fundamental declarative memories," says Justin Sanchez, who is in charge of neuroprosthetics research at the University of Miami in Florida.

Wired-up

Further developments should allow these neural chips to tackle more sophisticated problems than simple skills learning. "Think of the guy coming back from war who can't remember his wife's face," says Sanchez. For that kind of recognition, the brain breaks down the person, place or object into specific features – such as the colour of their hair or their height – and encodes them separately.

Using MIMO to replicate that process is an ambitious challenge that Deadwyler and Hampson have begun to explore more recently. For instance, they trained macaques to remember the position and shape of a picture on a screen, and then choose the same image from a much larger selection nearly a minute later. All the while the algorithm, via electrodes in the macaques' brains, recorded the neural signals that formed in the prefrontal cortex and hippocampus. Then they drugged the monkeys to disrupt their ability to form new long-term memories, before getting them to perform the task again. When they electrically stimulated neurons with the same signals that they had recorded on successful trials, those monkeys performed a lot better. By injecting the code, they had stimulated the hippocampus and the prefrontal cortex to reproduce the "correct" memories.

Intriguingly, Deadwyler and Hampson had found patterns that corresponded not to the exact images the monkeys were looking at, but to more universal features in them, such as whether they contained the colour blue, or a human face. "This is how we think memory works," says Deadwyler: instead of wastefully creating separate neural signatures for every new person, place or thing you encounter, the brain breaks the incoming information down into features. "Then to remember a specific item, you don't need to remember everything about it," he says. Rather than the fine details, it's the combination of features that helps bring the item in question back to mind.

The monkeys' own brain plasticity may have given the algorithm a helping hand, says Daofen Chen at the National Institutes of Health in Bethesda, Maryland. "The brain tries to meet the machine halfway," he says. "It is an adaptive process. Give the brain enough – even imperfect – information, and it can translate it into something it finds useful."

As the technology improves, brain chips that incorporate electrodes and algorithms like MIMO may be able to translate extremely fine details of an experience. Ranulfo Romo at the University of Mexico has shown that his chips can pick up the signals that capture very subtle changes in sensory perceptions, such as a certain frequency of vibrations against the skin. As a proof of principle, he even used the set-up to implant one monkey's ongoing sensations into another's brain, as if they were telepathic. "The monkeys integrated the false perception as their own working memory," says Romo.

Such fine-tuned decoding of sensory information could have important applications beyond restoring perceptual information to memories. For instance, sometimes people lose the ability to speak because of damage in the brain between Wernicke's area and Broca's area. A chip capable of picking up on

Electronic implants could help people with brain damage by recording neural signals and then relaying them around the damaged area

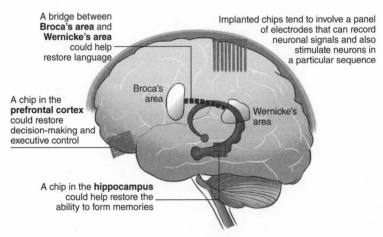

A bridge between **Broca's area** and **Wernicke's area** could help restore language

Implanted chips tend to involve a panel of electrodes that can record neuronal signals and also stimulate neurons in a particular sequence

A chip in the **prefrontal cortex** could restore decision-making and executive control

Broca's area

Wernicke's area

A chip in the **hippocampus** could help restore the ability to form memories

FIGURE 9.2 Rebuilding broken brains

those detailed sensory signals and translating them between the two regions might therefore return their speech (*see* Figure 9.2).

Despite these advances, the biggest unknown is the quality of the experiences. "For an animal, you can't ask them 'what is your perception of a memory?'" says Sanchez. That could soon change. The Restoring Active Memory project, run by Sanchez for the US Defense Advanced Research Projects Agency (DARPA), is pushing the research into human trials. Preliminary results suggest that the team involved has identified the electrical signals of simple memories as they form in the hippocampus of human volunteers and has been able to boost memory recall with a well-targeted electrical pulse.

After the competing algorithms and electronics have been tested and refined on volunteers, prototype chips will enter clinical trials. These studies will require the approval of the US

Food and Drug Administration and informed consent of the volunteers. If deemed sufficiently safe by the FDA, a chip will be cleared for use. DARPA hopes to use the resulting implants to help soldiers who return from war with traumatic brain injuries.

Several neuroengineering researchers envision similar chips for people with Alzheimer's disease and stroke, depending on the extent of the damage. In more severe cases of brain damage, Hampson imagines a device worn on the belt, with buttons that help you remember specific locations and their meanings. "Let's say I'm in the kitchen – I need to remember where the silverware is," he says. The patient would press the right button "and the memory pops up because we've stored the code".

Memory modification does, however, raise deeper questions than whether it can be done. As Luis Buñuel, arguably the father of surrealist film, put it: "Our memory is our coherence, our reason, our feeling, even our action. Without it, we are nothing." If you change those memories artificially, are you still you?

Deadwyler and Hampson's rat experiments highlight one possible concern: your memories may no longer be your own. "Activate the right circuits, and you generate the illusion that you are recalling something," says Romo. What kind of controls could ensure that every implanted memory reflected the reality of that person's environment? And whether or not those memories are your own, sparking neurons related to memory will eventually lead, directly or not, to changes in your decisions – so who is responsible for the consequences of those decisions?

There is also the chance that such chips could regurgitate long-buried events. Not all of those recollections will be wanted – one of the brain's biggest talents is to forget, as well as to remember. But perhaps it's a small price to pay for a lifetime of new memories to come.

The rise of the silicon brain

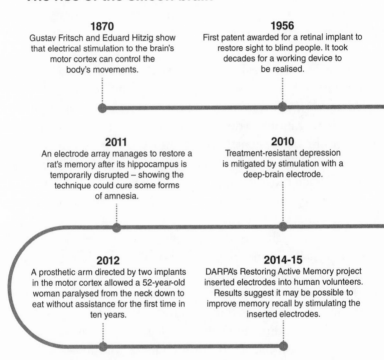

1870
Gustav Fritsch and Eduard Hitzig show that electrical stimulation to the brain's motor cortex can control the body's movements.

1956
First patent awarded for a retinal implant to restore sight to blind people. It took decades for a working device to be realised.

2011
An electrode array manages to restore a rat's memory after its hippocampus is temporarily disrupted – showing the technique could cure some forms of amnesia.

2010
Treatment-resistant depression is mitigated by stimulation with a deep-brain electrode.

2012
A prosthetic arm directed by two implants in the motor cortex allowed a 52-year-old woman paralysed from the neck down to eat without assistance for the first time in ten years.

2014-15
DARPA's Restoring Active Memory project inserted electrodes into human volunteers. Results suggest it may be possible to improve memory recall by stimulating the inserted electrodes.

1957
First human trial of cochlear implants, which transmit sound to the brain, showed that electronics can translate sensory information into the brain's language.

1996
Electrodes were implanted in the right and left hemispheres of a monkey's brain, giving it control of a prosthetic arm.

2008
Another volunteer with quadriplegia performs sophisticated movements with a prosthetic arm, controlled via a chip in his motor cortex that sent commands to the arm.

2004
A volunteer with quadriplegia tested BrainGate, a device implanted into his brain that allowed him to switch on lights and change channels on a television using only his thoughts.

10
Make the most of it

The human mind has all sorts of useful design features but also many glitches and weaknesses. The problem is, it doesn't come with a user manual. You just have to plug and play. Luckily, neuroscience can help with a few golden rules about how to get the best out of your brain.

The User's Guide to the Mind

Control your attention

Attention is the filter that the brain uses to decide what is important enough to be processed and acted upon. But there's a problem – distractibility comes as part of the package, and it is both a big and a design feature.

In simple terms, the brain has two attention systems. One, the "bottom-up" system, automatically snaps awareness to potentially important new information, such as moving objects, sudden noises or sensations of touch. This system is fast, unconscious and always on (at least when you are awake).

The other, the "top down" system, is deliberate, focused attention, which zooms in on whatever we need to think about and, hopefully, stays there long enough to get the job done. This is the form of attention that is useful for doing tasks that require concentration, but it is prone to running out of steam, and to being rudely interrupted by the bottom-up system.

One suggestion to stay focused is to cut down on bottom-up distractions by turning off email notifications, putting your phone on silent and so on. Another, according to Nilli Lavie, a cognitive neuroscientist at University College London, is to give your brain more to do.

Lavie's work has shown that better control of top-down attention comes not by reducing the number of inputs, but by increasing them. Her load theory says that once the brain reaches its limit of sensory processing, it can't take anything else in, including distractions.

So it might help to make a task more attention-grabbing by putting a colourful border around a blank document and making the bit you are working on purple, for example. Lavie says that it works with all the senses, so choosing somewhere with a bit of background noise might also help.

Early work suggests that cognitive training might help, as could meditation. Long-term meditators have been shown to have thicker parts of the brain associated with attention, while other studies have found that attention test scores improved after a short course of meditation. So learning to focus better may be as simple as making time to sit still and focus on not very much.

Boost memory

Like attention, working memory is one of the brain's most crucial front-line functions. Everything you know and remember, whether it's an event, a skill or a fascinating fact, started its journey into storage by going through your working memory (*see* Chapter 1 for more on this).

Working memory comes as standard in the human brain, but it is well worth trying to maximise what you've got. Working memory capacity is a better predictor of academic success than IQ.

The good news is that the system can probably be upgraded. Some studies have shown that brain training programmes aimed specifically at working memory can produce improvements, and there are even a handful of training packages on the market. But it's not clear whether they make you better at anything other than working memory tests.

Cognitive neuroscientist Jason Chein of Temple University in Philadelphia, Pennsylvania, who studies working memory, says there seems to be evidence of improvements in other cognitive skills, although any changes are quite small. "A small effect may still be important in the sense that even modest gains can have a meaningful impact on everyday cognition," he says.

Learn like a child

Learning is what your brain does naturally. In fact, it has been doing it every waking minute since about a month before you were born. It is the process by which you acquire and store useful (and useless) information and skills. Can you make it more efficient?

FIGURE 10.1 It's no surprise that focused attention is the best way to retain a fact

The answer lies in what happens physically as we learn. As it processes information, the brain makes and breaks connections, growing and strengthening the synapses that connect neurons to their neighbours, or shrinking them back. When we are actively learning, the making of new connections outweighs the breaking of old ones. Studies in rats have shown that this rewiring process can happen very quickly – within hours of learning a skill such as reaching through a hole to get a food reward. And in some parts of the brain, notably the hippocampus, the brain grows new brain cells as it learns.

But once a circuit is in place, it needs to be used if it is going to stick. This largely comes down to **myelination** – the process whereby a circuit that is stimulated enough times grows a coat of fatty membrane. This membrane increases conduction speed, making the circuit work more efficiently.

What, then, is the best way to learn things and retain them? The answer won't come as a huge surprise to anyone who has been to school: focus attention, engage working memory and then, a bit later, actively try to recall it.

Alan Baddeley of the University of York, UK, says it is a good idea to test yourself in this way as it causes your brain to strengthen the new connection. He also suggests consciously trying to link new bits of information to what you already know. That makes the connection more stable in the brain and less likely to waste away through underuse.

The learning process carries on for life, so why is it so much harder to learn when we reach adulthood? The good news is that there seems to be no physiological reason for the slowdown. Instead, it seems to be a lot to do with the fact that we simply spend less time learning new stuff, and when we do, we don't do it with the same potent mix of enthusiasm and attention as the average child.

Part of the problem seems to be that adults know too much. Research by Gabriele Wulf at the University of Nevada, Las Vegas, has shown that adults tend to learn a physical skill, like hitting a golf ball, by focusing on the details of the movement. Children, however, don't sweat the details, but experiment in getting the ball to go where they want. When Wulf taught adults to learn more like kids, they picked up skills much faster.

This also seems to be true for learning information. As adults we have a vast store of mental shortcuts that allow us to skip over details. But we still have the capacity to learn new things in the same way as children, which suggests that if we could resist the temptation to cut corners, we would probably learn a lot more.

A more tried-and-tested method is to keep active. Ageing leads to the loss of brain tissue, but this may have a lot to do with how little we hare about compared with youngsters. With a little exercise, the brain can spring back to life. In one study, 40 minutes of exercise three times a week for a year increased the size of the hippocampus – which is crucial for learning and memory. It also improved connectivity across the brain, making it easier for new things to stick.

Can the brain run out of storage space?

The good news is that there seems to be no limit to the knowledge that can fit into a brain. As far as we know no one has ever run out of storage space. But it seems you can know too much. Michael Ramscar at Tübingen University in Germany reckons that anyone who lives long enough eventually hits that point just by virtue of a lifetime's knowledge. He suggests that cognitive skills slow down with age not because the brain withers but because it is so full. And that – like an overused hard drive – takes longer to sift through.

Get creative

J. K. Rowling has said that the idea for Harry Potter popped into her head while she was stuck on a very delayed train. We have all had similar – although probably less lucrative – "aha" moments, where a flash of inspiration comes along out of the blue. Where do they come from? And is there any way to order them on demand?

Experiments led by John Kounios, a neuroscientist at Drexel University in Philadelphia, suggest that the reason we aren't all millionaire authors is that some brains come better set up for creativity than others. EEG measurements taken while people were thinking about nothing in particular revealed naturally higher levels of right hemisphere activity in the temporal lobes of people who solved problems using insight rather than logic. Kounios says recent work hints that this brain feature might be inherited, but even if you happen to have a more focused, less creative brain, there are plenty of general tips on how to get it into creative mode.

Boringly, the first is to put in the groundwork to build up a good store of information so that the unconscious has something to work with. Studies on subliminal learning have poured cold water on the idea that knowledge can drift into the brain without any conscious effort, so it pays to focus intently on the details of the problem until all the facts are safely stored. At this stage, anything that helps with focus, such as caffeine, should help.

Once that's taken care of, it's time to cultivate a more relaxed, positive mood by taking a break to do something completely different – like watching a few entertaining cat videos. Studies where people have either watched a comedy film or a thriller before coming up with new ideas have shown that a relaxed and happy mood is far more conducive to ideas than a tense and anxious one. Not only that, but it pays to turn down the focus knob a little, and the easiest way to do that is to look for ideas when your brain is too tired to focus properly.

A 2011 study showed that morning people had their most creative ideas late at night, while night owls had theirs early in the morning. Mental exhaustion might be a more realistic state of mind than relaxation when an important deadline is looming, but if the ideas are still refusing to come there may one day be an easier solution. Brain stimulation studies, in which activity was boosted in the right temporal lobe and suppressed in the left, increased the rate of problem-solving by 40 per cent. So the stressed creative of the future might be able to pop on a "thinking cap" to help those juices flow.

Boost intelligence

Intelligence has always been tricky to quantify, not least because it seems to involve most of the brain and so is almost certainly not one "thing". Even so, scores across different kinds of IQ tests have long shown that people who do particularly well – or badly – on one seem to do similarly on all. This can be crunched into a single general intelligence factor, or "g", which correlates pretty well with academic success, income, health and lifespan, and seems to be largely genetic (*see also* chapter on Intelligence).

That doesn't mean the environment plays no part, at least in childhood. While the brain is developing, everything from diet to education and stimulation plays a huge part in developing the brain structures needed for intelligent thought. Children with a bad diet and poor education may never fulfil their genetic potential.

But even for educated and well-fed children, the effects of environment wear off over time. By adulthood, genes account for 60 to 80 per cent of the variance in intelligence scores, compared with less than 30 per cent in young children. Like it or not, we get more like our close family members the older we get.

The good news is that one type of intelligence keeps on improving throughout life. Most researchers distinguish between fluid intelligence, which measures the ability to reason, learn and spot patterns, and crystallised intelligence, the sum of all our knowledge so far. Fluid intelligence slows down with age, but crystallised intelligence doesn't. So while we all get a little slower as we age, we can rest assured that we are still getting cleverer.

Time it right

The brain is a fickle beast – at some times as sharp as a tack, at others like a fuzzy ball of wool. At least some of that variation can be explained by fluctuations in circadian rhythms, which means that, in theory, if you do the right kind of task at the right time of day, life should run a little more smoothly.

The exact timing of these fluctuations varies by about two hours between morning and evening types, so it is difficult to give any one-size-fits-all advice. Nevertheless, there are a few rules that it's worth bearing in mind whatever your natural waking time.

It's an idea not to do too much that involves razor-sharp focus in the first couple of hours after waking up. Depending on how much sleep you have had it can take anything from 30 minutes to 4 hours to shake off sleep inertia – also known as morning grogginess. If you want to think creatively, though, groggy can be good.

If hard work can't wait, though, the good news is that researchers have backed up what most of us already know – a dose of caffeine helps you shake off sleep inertia and get on with some work.

Another tip is to time your mental gymnastics to coincide with fluctuations in body temperature. Studies measuring variations in everything from attention and verbal reasoning to reaction times have shown that when our core temperature dips below 37°C the brain isn't at its best.

By this measure, the worst time to do anything involving thinking is, unsurprisingly, between midnight and 6am. It is almost as bad in the afternoon slump between 2pm and 4pm, which has more to do with body temperature than lunch – studies of people who have no lunch or just a small one have the same problem. All in all, the best time to get stuck in is between mid-morning and noon and then again between 4pm and 10pm.

There may be a way to hack the system, though. Studies have shown that body temperature changes and alertness also work independently of the internal clock, so a well-timed bit of exercise or hot shower can work wonders.

Competitive sports, though, are worth leaving until the end of the day. Studies have shown that reaction times and hand-eye coordination get progressively better throughout the day, reaching a peak at around 8pm.

After that, there's time for a little more focused energy before the body cools down, the brain slows and gets ready to drop off and prepare for another day.

Interview: The woman who rebuilt her brain

The adult brain is less malleable than a child's, but it continues to change throughout our entire lives. After a stroke or brain injury the brain is particularly plastic, pulling out all the stops to get around the roadblock.

Neuroscientist **Jill Bole Taylor** *suffered a stroke in 1996, which robbed her of her memory, motor skills and even personality. During her eight-year recovery she put her knowledge of the brain to good use to help rebuild her mind to be better than before.*

What did the stroke do to you?

Because the haemorrhage was in my cerebral cortex, it wiped out my cognitive mind. I was very fortunate, though, in that my body was going to be OK.

Describe the days that followed.

I was in hospital for five days. On the morning of the third day my mother came to my side. Now, I did not know what a mother was, much less who *my* mother was. She came in, acknowledged everyone in the room, and then immediately picked up the sheet and crawled into bed with me. I didn't know who this person was. I didn't know what this person was. All I knew was that this very kind woman just crawled into my bed, wrapped her arms around me and started rocking me, like I was her baby. And I was her baby. She just recognised that I was an infant again and that was that.

What did you do for your rehabilitation?

The only formal rehab I had was speech therapy. I saw a speech therapist for about three months. My real rehab was done by my mother from the day she brought me home. She was an angel in my life. She would take me to the bathroom, feed me and then if I had any energy left she would work me — children's puzzles, teaching me to read, walking me around the apartment and then the block, those kinds of things. I would not be here if it were not for her.

The advantage I had was that I believed in the ability of the brain to recover itself. That meant primarily for me to get out of its way.

How do you get out of the brain's way?

My number one recommendation is sleep. The brain needs sleep. These cells have been traumatised. The person is totally burned out and fried, and they want to sleep. In our society, generally what happens in a rehabilitation environment is that wake-up time is at 7am. Everyone gets awakened.

If you are a stroke survivor and you are zoned out and don't want to be awake, you will be pumped with amphetamines. Stimulus is stuck in your face, often in the form of a TV set in the room, sometimes literally a foot from your face. It's pure pain. And then we keep these people awake through dinner. After dinner they're put back to bed. The idea is that if you're going to recover, you have to act like a normal person. If that had been my experience, honestly I would have chosen not to engage. There's no question in my mind that we're not treating stroke survivors effectively.

You have said that you retreated into the right hemisphere of the brain. What was that like?

When I had the haemorrhage, the personality of my left hemisphere was traumatised. I shifted all the way into the right hemisphere, because the left-brain personality became non-functional and released her dominance, or released the dominating neural fibres that were inhibiting my right hemisphere. That's from an anatomical perspective.

As time went on, different circuits in the left hemisphere started to become functional again. It was like repairs. So it was a long process of me in relationship with my brain, day after day, year after year, rebuilding. I was consciously choosing and rebuilding my brain to be what I wanted it to be.

Did you actually consciously reconstruct your brain with your thoughts?

Yes, renewing or rerunning neurocircuits was a cognitive choice. The non-functional circuits started to come back online one at a time and I could choose to either hook into that circuitry or not feed it. For example, when the anger circuit wanted to run again, I did not like the way it felt inside my body so I said "no" to its running. Every time it tried to get triggered and run again, I brought my attention back to it – I did not like the way anger felt so I shut it down. Now that circuit rarely runs at all, mostly because I feel it getting triggered and nip it in the bud.

"When the circuits came back, I could choose to hook into them or not"

It was so clear to me during my recovery that every ability I had was because the circuit that controlled it was good, it was functioning. I learned that certain thoughts that I had could stimulate the emotional circuitry, which could then result in a physiological response.

So, I look at us as a collection of neurocircuitry of thoughts and emotions and physiological responses. When you see the brain as the kind of computer network that it is, it becomes easier to manipulate. But you have to be willing. People say "Oh I'm so much more than my thoughts, I'm so much more than neurocircuitry," and I'm like, yeah, I had that fantasy once, too. I don't any more. As human beings we all have the ability to focus our minds on what we want to think about.

This sounds like the claims made by meditators.

I think folks who meditate are willing to pay attention to their thoughts so that they can purposefully redirect their minds. Mantras, prayer beads, consciously thinking about one's breathing – these are tools that provide the brain with an alternative to the constant brain chatter, permitting the mind's focus to shift to something else. It's the same sort of thing. Learning to observe our neural circuitry and not engage with it is a skill we all can learn.

When did you know you had recovered?

I felt I was completely recovered when I felt I had become a solid again. Up until then I felt that I was a fluid. I'd get up in the morning and take my dog out. I have woods out back, and I knew I had recovered when everything blended, everything radiated the energy of life – the trees and the light coming through them, the grass and the sparkling dew. Everything was vivid, beautiful and connected, and I was a part of it all. That's very different to saying "I am a solid, and that's a tree and that's a blade of grass and that's a drop of dew," and everything is separate. I don't know how else to describe it.

You do a lot of stained-glass work now. Has your perception of the artwork, and indeed your life, changed much since your stroke?

Oh yeah, everything's more vibrant, more alive and more beautiful now. More fluid, more curves, fewer lines, more relative, less disconnected, more similar, less different. Everything in my life has changed like that since the stroke. If someone said to me, "Okay Jill, we're going to put you in a time capsule and let you wake up that day again and you get to choose to have the stroke or not have it," I would have the stroke in a minute.

Jill Bolte Taylor is a neuroanatomist affiliated with the Indiana University School of Medicine and author of *My Stroke of Insight*.

Change your brain the easy way

Boosting your mental faculties doesn't have to mean studying hard or becoming a reclusive bookworm. Here are some tried and tested methods – minimal effort required.

Food for thought

Eat breakfast to give blood sugar levels an early boost. Preferably high in slow-release carbs and with as many vitamins as possible – particularly B vitamins.

A varied diet is best, but these few have particular reasons to be considered brain foods:

- **Eggs**, rich in choline, which the body uses to produce the neurotransmitter acetylcholine (crucial for memory).
- **Yogurt** contains the amino acid tyrosine, needed for the production of the neurotransmitters dopamine and noradrenalin, among others.
- Brains are around 60 per cent fat, but trans-fats seem to clog up the system. Instead, aim for omega-3 fatty acids, in particular docosahexaenoic acid or DHA found in large quantities in **oily fish**.
- **Blueberries and other berries** – higher intake is linked with improved cognitive skills.

Sleep on it

Sleep deprivation is the enemy of planning, problem-solving, memory and IQ. Catching up removes the problem, but beware of getting too much (*see* Chapter 8). Seven hours seems to be the magic number for a brain that functions as well as it can.

Exercise

Walking sedately for half an hour three times a week can improve abilities such as learning, concentration and abstract reasoning by 15 per cent. Exercise not only gets blood flowing, it also boosts the growth of new brain cells in the hippocampus, potentially making more room for new memories. Weirdly, the effect seems to work both ways: mental exercise can tone up the body. In 2001, researchers at the Cleveland Clinic Foundation in Ohio asked volunteers to spend just 15 minutes a day thinking about exercising their biceps. After 12 weeks, their arms were 13 per cent stronger.

Smart drugs

According to a recent survey, 38 per cent of those 750 people surveyed had taken a cognitive enhancing drug. Many had bought them online. Research has linked the most common, modafinil – a narcolepsy drug – to stay awake for long periods without the side-effects of stimulants like caffeine and amphetamines and with no need to catch up on lost sleep. Similarly, many people are using Ritalin not because they suffer from attention deficit disorder, but because they want to concentrate better to study or work. So far it isn't clear if there are any long-term side effects that we should be worried about from all of this.

The pharmaceutical pipeline is clogged with promising compounds specifically designed to augment memory, so perhaps a safe, all-purpose brain-boosting pill is on the way. Watch this space.

Fifty ideas

This section helps you to explore the subject in greater depth, with more than just the usual reading list.

Five ideas for places to visit

1 Ramon y Cajal – Library of the Cajal Institute in Madrid, where there is a small exhibition of Cajal's game-changing drawings of neurons dating from the early 1900s. By prior appointment only.
Contact details at: http://www.cajal.csic.es/ingles/

2 See the skull of Phineas Gage, and the tamping iron that went through it, destroying his left frontal lobe. Gage's freak injury in 1848 revealed the frontal lobe's role in personality and impulse control.
Warren Anatomical Museum, Harvard, 10 Shattuck St. Boston, MA 02115
https://legacy.countway.harvard.edu/menuNavigation/chom/warren/exhibits.html

3 Einstein's brain can be found at the Mütter Museum, 19 S 22nd St, Philadelphia, PA 19103 http://muttermuseum. org/exhibitions/albert-einsteins-brain/

4 Freud's House, London. See the famous couch where Freud attempted to delve into the dark recesses of his clients' minds. 20 Maresfield Gardens, London NW3 5SX https://www.freud.org.uk/

5 Phrenology heads – Before we were able to look inside the brain, phrenologists tried to understand behaviour via the bumps on our heads and the shape of our skulls. The archive of the Edinburgh Phrenological Society, including variously shaped heads, is on display at the Anatomical Museum, Edinburgh University.
http://www.ed.ac.uk/biomedical-sciences/anatomy/anatomymuseum/exhibits/masks

Ten facts

1 Size doesn't necessarily matter. An average male brain weighs around 1,400g. Einstein's brain, at 1,230g, was slightly below average. Anatole France, who won the Nobel Prize in Literature in 1921 had even less up top at 1,017g. Whatever made them geniuses, it certainly wasn't a larger-than-average brain.

2 Pain is perceived in the brain but the brain itself contains no pain receptors. This is how some kinds of brain surgery can be done while the patient is awake. The blood vessels and membranes around the brain, however, do have pain receptors, which is why we get headaches.

3 While in the womb, up to a quarter of a million new cells form every minute, making 1.8 million new connections per second. About half of these will later be lost, leaving only those reinforced by use.

4 If you unravelled the human cerebral cortex it would measure 2,500cm², about the same as an A2 size sheet of paper.

5 The average person spends 25 years of their life, around a third of the average lifespan, asleep.

6 The idea that we use only 10 per cent of our brains is a myth. We use all of it, but not all at once.

7 Neurons are not the most common cells in the brain. This prize goes to the **glia**, which provide structure and technical support to the neurons.

8 Neurons that have been covered in **myelin**, a fatty white sheath that is added to the most commonly used pathways in the brain, can transmit electrical messages ten times faster than an unmyelinated neuron.

9 The magnet in an MRI scanner is as powerful as the ones used to pick up cars in a scrapyard.

10 An Iron Age brain, still preserved inside the skull of a man killed 2,600 years ago, was discovered in Yorkshire in 2008. Brains usually liquefy soon after death so this discovery, in oxygen deprived, waterlogged soil, was incredibly rare.

Ten quotes

1 "Men ought to know that from the brain, and from the brain only, arise our pleasures, joy, laughter and jests, as well as our sorrows, pains, griefs, and tears."
Hippocrates (about 400 BCE).

2 "As long as our brain is a mystery, the universe, the reflection of the structure of the brain will also be a mystery."
Santiago Ramon y Cajal.

3 "People who boast about their I.Q. are losers."
Stephen Hawking.

4 "The highest activities of consciousness have their origins in physical occurrences of the brain, just as the loveliest melodies are not too sublime to be expressed by notes." W. Somerset Maugham.

5 "All that we know, all that we are, comes from the way our neurons are connected."
Sir Tim Berners-Lee.

6 "There is no scientific study more vital to man than the study of his own brain. Our entire view of the universe depends on it."
Francis Crick.

7 "Any man who reads too much and uses his own brain too little falls into lazy habits of thinking."
Albert Einstein.

8 "If I had to live my life again I would have made a rule to read some poetry and listen to some music at least

once a week; for perhaps the parts of my brain now atrophied could thus have been kept active through use." Charles Darwin.

9 "Until you make the unconscious conscious, it will direct your life and you will call it fate." Carl Jung.

10 "Toleration is the greatest gift of the mind; it requires the same effort of the brain that it takes to balance one-self on a bicycle." Helen Keller.

Five literary references

1 "I am a brain, Watson. The rest of me is a mere appendix."
   Arthur Conan Doyle's Sherlock Holmes in *The Adventure of the Mazarin Stone.*

2 "My own brain is to me the most unaccountable of machinery – always buzzing, humming, soaring roaring diving, and then buried in mud. And why? What's this passion for?"
   Virginia Woolf.

3 "If both the past and the external world exist only in the mind, and if the mind itself is controllable – what then?"
   George Orwell, *1984.*

4 "Of what a strange nature is knowledge! It clings to a mind when it has once seized on it like a lichen on a rock."
   Mary Shelley, *Frankenstein.*

5 "I like nonsense, it wakes up the brain cells."
   Dr Seuss.

Five jokes

1 What is everyone's favourite band when they're asleep?
REM

2 Where would your brain choose to go on holiday?
Hippocamping

3 What do you call a head *without* 86 billion neurons?
A no-brainer

4 I finally figured out what was wrong with my brain. On the left there's nothing right and on the right there's nothing left.

5 When should you carry a big umbrella?
During a brainstorm.

For more brain jokes, *see*:
http://faculty.washington.edu/chudler/jokes.html

Five things for kids

1 **Brain jelly**: Search online for a brain-shaped jelly mould. Turn it upside down and fill with a tangle of strawberry laces. Add light pink jelly or blancmange and leave to set. Makes a treat for all brain enthusiasts – and great for Halloween.

2 **Test your reaction time**: One person holds a 30cm ruler at the 30cm end and lets it hang down. The other person puts their hand beneath the ruler, ready to catch it when the first person drops it. Person 1 drops the ruler within 5 seconds. Note the number on the ruler where person 2 catches it. Repeat 3 times and take an average. This can be converted to reaction time as below.

Distance	Time
2 in (~5 cm)	0.10 sec (100 ms)
4 in (~10 cm)	0.14 sec (140 ms)
6 in (~15 cm)	0.17 sec (170 ms)
8 in (~20 cm)	0.20 sec (200 ms)
10 in (~25.5 cm)	0.23 sec (230 ms)
12 in (~30.5 cm)	0.25 sec (250 ms)

3 **The memory tray**: Add a handful of small objects to a tray. Show the tray to the child for 30 seconds then remove it and ask them to recall as many items as they can. Repeat with the whole family to see whose memory is best.

4 **A smartphone-based reading game** developed by
researchers in Finland, which aims to help children
with little access to education escape poverty by
becoming literate.
http://info.graphogame.com/

5 **Visual illusions online:** A treasure trove of visual
illusions and kid-friendly explanations
http://psylux.psych.tu-dresden.de/i1/kaw/diverses%20
Material/www.illusionworks.com/

Ten ideas for finding out more

1 BrainFacts.org
A public outreach website from the Society for Neuroscience.

2 Neuroscience for Kids. Fun neuroscience facts and explainers from University of Washington neuroscientist Eric Chudler.
http://faculty.washington.edu/chudler/neurok.html

3 *The Brain: A very short introduction* by Michael O'Shea. Oxford University Press (2005).

4 *NeuroPod*: The neuroscience podcast of *Nature* magazine.
http://www.nature.com/neurosci/neuropod

5 The whole brain atlas: a collection of brain scans in health and disease.
http://www.med.harvard.edu/AANLIB/home.html

6 The brain from top to bottom.
A web-based introduction to the brain by researchers at McGill University in Montreal, Canada.
http://thebrain.mcgill.ca/avance.php

7 The Human Connectome Project aims to map the brain's circuit diagram. See latest project updates here.
https://www.humanconnectome.org/

8 *The Brain with David Eagleman.* A six-part television series available on DVD and with an accompanying book.

9 Testmybrain.org
Get a measure of your brain skills compared to others
and contribute to neuroscience studies while you are at it.

10 An interactive introduction to neuroscience from the
UK-based science centre @Bristol.
http://www.youramazingbrain.org.uk/

Picture credits

Figure 1.1 Structure of a neuron

Figure 1.2 Identifying the major parts of the brain

Figure 1.3 Penfield's homunculus: How the brain sees the body © Natural History Museum, London/Science Photo Library

Figure 1.4 The brain's 12 "rich-club" hubs

Figure 1.5 Size isn't everything

Figure 2.1 Memory: from sense to storage

Figure 2.2 A single concept may be represented by 1 million neurons in your hippocampus (though these belong to a sea slug) © B. M. Salzberg; D. Kleinfeld; T. D. Parsons; and A. L. Obaid

Figure 2.3 You must remember this

Figure 2.4 Testing the ability to suppress memories

Figure 3.1 Alfred Binet devised the first IQ test © Universal History Archive/Rex/Shutterstock

Figure 3.2 Average IQ score distribution by population

Figure 3.3 The "three stratum theory" of intelligence

Figure 3.4 What makes someone smart?

Figure 3.5 The pattern of IQ scores among military conscripts in Norway

Figure 4.1 An expressive Serena Williams © BPI/BPI/Rex/Shutterstock

Figure 4.2 Emotional expressions

Figure 5.1 Cortical sensory areas

Figure 5.2 Troxler's fading

Figure 5.3 Split-second timing relies on the brain's ability to predict which perceptual stimuli belong together © Tim de Waele/Corbis via Getty

Figure 5.4 The now illusion

Figure 6.1 Seats of consciousness

Figure 6.2 In theory we could calculate how conscious anything is, be it human, rat or computer © Cultura/Rex/Shutterstock

Figure 6.3 Who's in charge? An experiment on free will

Figure 7.1 The adolescent stage of brain development *Source*: PNAS, Vol. 101. Copyright (2004) US NAS

Figure 7.2 The decline in cognitive ability *Source*: *American Journal of Psychiatry*. Vol. 156 p.58

Figure 7.3 Looks like Venus and Mars aren't so far apart after all © VISUM/Rex/Shutterstock

Figure 8.1 Human slumber cycles

Figure 8.2 Why do babies spend the majority of their time asleep? © WestEnd61/Rex/Shutterstock

Figure 8.3 How much shuteye do I need? Sleep need varies as we age, and for each individual.
Source: National Sleep Foundation

Figure 9.1 Thought-controlled wheelchair © Philippe Psaila/ Science Photo Library

Figure 9.2 Rebuilding broken brains

Figure 10.1 It's no surprise that focused attention is the best way to retain a fact © Bryn Colton/Getty

Index